Urban Infrastructure: Reflections for 2100

Urban Infrastructure: Reflections for 2100

Edited by Sybil Derrible and Mikhail Chester
with contributions from 40 global infrastructure experts

2020

2020, 1st. Edition.

ISBN 9798695826524

Edited by: *Sybil Derrible & Mikhail Chester*
Cover design: *Callie W. Babbitt & Nora C. Babbitt*
Visit our website 2100.resilientinfrastructure.org.

To future infrastructure leaders
who will have to navigate
a remarkably more
complex world.

Contents

— Preface —

From food, water, and shelter, to entertainment and education, *infrastructure* represent the cornerstone of human activity. While providing unquantifiable benefits to humanity, infrastructure are caught between the past and the future. Once built, infrastructure immediately become manifestations of the technologies and needs from their past. In an effort to improve their efficiency and augment their capabilities, infrastructure are becoming increasingly *hybridized*, joined with emerging sensing, analytical, and automated cyber technologies. Yet, it remains unclear how infrastructure needs to evolve for a world that is rapidly changing. What should infrastructure *do* at the end of the century? How should infrastructure systems be *structured*, physically and institutionally? How do we identify and correct mistakes from the past?

The twentieth century has changed the world forever. From 1.6 billion people in 1900, the world population increased to 6.1 billion in 2000. Beyond this exponential growth in population, the consumption of virtually everything has also grown exponentially, including energy, fertilizer, water, mobility, and telecommunications, to name a few. This growth was in part driven by infrastructure investments. The twentieth century saw the rise of the automobile, the genesis of the highway, the development of water distribution systems in cities around the globe, the creation of rigorous wastewater collection and solid waste management practices, the expansion of electricity to virtually every household, and the birth of the internet. By 2100, the world population is expected to near 11 billion people, be connected with cybertechnologies and artificial intelligence that we cannot fathom, be subject to major changes in earth systems including water, nutrients, and climate, and possibly be driven by radically different needs and goals than those of the twentieth century. What sort of infrastructure evolution will be required for this population given these new constraints?

Indications are that infrastructure in the twenty-first century will need to embrace sustainability, resilience, complexity, and cyber-technologies, while at the same time supporting technologies and processes that correct the negative impacts of the last century. This is a tall order. How can we even start to try to shift infrastructure for the future?

Italo Calvino in his 1972 book *Invisible Cities* positions explorer Marco Polo with Emperor Kublai Kan to describe the infrastructure, cultures, languages, and human experiences in cities across the growing Mongolian Empire. There is nothing scientific about these accounts (and most if not all cities are fictitious) but they enable the reader to travel and experience the city being described. The approach for the book is closer to science fiction than an accurate historical account, and it works surprisingly well. The book was gifted to Prof. Derrible by Prof. Chester and got us thinking about how to create a space where leading infrastructure thinkers can *creatively* explore pathways towards future infrastructure. The result was this book, for which we have invited selected experts to contribute a *reflection* on infrastructure for 2100.

A *reflection* could be virtually anything; from an opinion piece or a short story to a poem or a drawing, or just about anything a contributor wishes to submit. In academia, we are conditioned to publish rigorous, peer-reviewed scientific work. For a reflection, the emphasis was shifted from rigor and accuracy to creativity and playfulness. Instead of coming up with a scientific method and sophisticated mathematical models to predict what infrastructure will be like in 2100, experts were invited to express their intuitions and feelings. This approach aligns with a quote from the theoretical physicist and mathematician Freeman Dyson who wrote in his 1988 book *Infinite in All Directions:* "There are two ways to predict the progress of technology. One way is economic forecasting, the other way is science fiction. Economic forecasting makes predictions by extrapolating curves of growth from the past into the future. Science fiction makes a wild guess and leaves the judgment of its plausibility to the reader... For the future beyond ten years ahead, science fiction is a more useful guide than forecasting."

In this book, the reader will find a collection of reflections. Some will directly discuss infrastructure in 2100, others will not. Some will carry positive messages and others will carry negative or mixed messages. All reflections were written in 2019 and 2020; two years that proved to be complicated in the history of mankind: the COVID-19 pandemic, climate change and extreme weather events, and the rise of authoritarian governments across the globe, all signs that manifest a need for systemic change. Realistically, we cannot predict how infrastructure will be in 2100, but the reflections in the book provide dialogue and visions of what we might hope for or fear, and only time will tell as to which side of the balance we end up leaning.

— Acknowledgements —

This work was in part supported by several U.S. National Science Foundation grants including the Urban Resilience to Extremes Sustainability Research Network (Award No. SRN-1444755), Converging Social, Ecological, and Technological Infrastructure Systems (SETS) for Urban Resilience project (Award No. GCR-1934933), and the Understanding the Fundamental Principles Driving Household Energy and Resource Consumption project (Award No. CAREER-1551731).

The editors thank the Infrastructure Misfits [un]society for their spirit determination towards transforming infrastructure for the Anthropocene. Additionally, they are appreciative of the work by Jennifer Jones who helped organize and edit the book and of Clemens Lode (www.lode.de) for his help compiling the book into digital format.

Editor Derrible thanks Marie-Agathe Simonetti for her enthusiasm and patience. As an art historian, Simonetti has opened Derrible to a more beautiful and creative world. Derrible would also like to thank the University of Transport Technology in Hanoi (Vietnam) for providing support during his sabbatical.

Editor Chester thanks Becca, Ethan and Abbie for their support and patience. This book hopefully serves as a small step towards making the future a brighter place. He also thanks Arizona State University and the University of California Los Angeles (Institute of Transportation Studies and Department of Civil and Environmental Engineering) for his sabbatical support and resources.

Foremost, the editors thank the authors for their enthusiasm, contributions, and willingness to be creative. The collective voice is stronger than the sum of the parts.

— Prologue —

Braden Allenby[†]

We live in the first chapter of a science fiction book. That is what you don't understand. That is why every time you try to make sense of your world, and maybe even think you may be succeeding, you fail. That is why you fall back, finally defeated, into your core zeitgeist, be it fundamentalist religion, nationalism, political tribe, identity brand, or anti-vaxx. What kind of scifi? – Happy? Noir? Gibson or Stephenson internet play world? Virtual Disney? No one knows yet. But that doesn't mean we can't try to guide how the scifi reads. But for that we need new tools, new approaches. Hence, this book.

In more stable times, books by engineers about infrastructure and the built environment are pretty straightforward. They contain recipes, and tables, and algorithms, and practices . . . all validated by past experience, and known to produce effective results that work IRL – in the real world. But none of those books recognize the fundamental assumption they all must make to be useful, and provide solutions to real problems – the thing that engineers, and technologists, and their managers, are paid to do. And that assumption is stability. And a particular kind of stability to boot: it is stability in the rate of change, so that the cycle times of all the multidimensional domains that must intersect for something to work IRL – from technology to institutions to communities to politics, law and regulation – are connected. No one part of the universe around infrastructure changes so fast that it decouples from all the other parts, so that evolution, as it occurs, is orderly and allows for integrated adaptation to change. This is the Ecclesiastes world:

One generation passeth away, and another generation cometh: but the earth abideth for ever. . . . All the rivers run into the sea; yet the sea is not full; unto the place from whence the rivers come, thither they return again. . . . The thing that hath been, it is that which shall be; and that which is done is that which shall be done: and there is no new thing under the sun.

[†]Arizona State University | brad.allenby@asu.edu

But as I write this in 2020, we do not live in such a world. We have decades old transportation infrastructure, and a complex financial, technological, and cultural infrastructure that makes gasoline and distributes it around the world – just as cars, trucks, and planes become autonomous and electric. The Federal Food and Drug Administration is running the MILES – metformin in longevity study – to see if a common drug used to treat diabetes is the first anti-aging drug; many researchers are confident that the first people to live to 150 with a high quality of life have already been born in developed countries. And they whisper soto voice that, if they can keep you alive until 150 . . . they can give you virtual biological immortality. No one has thought about resource demand or inequality or infrastructure evolution in such a long life world. AI, big data, and social media are creating new technological opportunities for those who understand behavioral economics to manipulate target populations without the individuals involved knowing that they are being manipulated, with implications that include the possible collapse of pluralistic governance systems. Private firms are developing mesh technology that can be injected into your skull, settle around your brain, and give you "chip in the brain" cognitive functionality, which might include such services as technologically enabled telepathy . . . not to mention brain hacking at a whole new level. Factory meat - meat produced from single cells in factories rather than in animals – is scheduled to be commercially available by 2022; if it is inexpensive enough, it might dramatically affect climate change, the nitrogen, phosphorous, and water cycles, global economic patterns, and the ability of poor countries to develop – among other things. And these are only a few scattered examples, because foundational, unpredictable, and accelerating change is occurring across the entire technological frontier.

And, of course, it isn't just technology. Geopolitics are morphing rapidly: the Western universalism world view, and the Westphalian world order it spawned, are failing in many places as Russia, China, Islam, and many other entities reject it in whole or in part. Democracy, which only a few decades ago seemed the logical choice for a new and modern world, is suddenly in crisis, from Brexit to Trumpian America, and China, bolstered by a recent history of amazing economic growth and a social credit system powered by vast data collection capabilities and state-level AI, becomes a viable soft authoritarian alternative. Core values such as free speech segue from legal systems to the terms and conditions of the big social media companies: if you can't get on Facebook, Twitter, WeChat, TikTok, and Tencent, you can scream all you want out your window, but no one will care. In such a world, cycle times of change have decoupled.

Technology changes too fast for humans and their institutions to adapt; the law and regulation fall further and further behind the reality that they are meant to govern. Tribalism trumps social cohesion as community narratives fail and local narratives become critical sources of identity, a process that often reflects subtle disinformation campaigns by domestic and foreign interests. Chaos is not unique to our age – remember Yeats' words in "The Second Coming," written in 1919:

Things fall apart; the centre cannot hold; Mere anarchy is loosed upon the world, The blood-dimmed tide is loosed, and everywhere The ceremony of innocence is drowned; The best lack all conviction, while the worst Are full of passionate intensity. But we are clearly in Second Coming space, not Ecclesiastes space, and such times do require a different kind of professional capability.

In particular, when the rate of change is steady and can be integrated in existing mental models and institutions, it is relatively easy to adjust in real time to even random events. But in Second Coming space, trying to adjust requires significantly expanding one's perspective to include what would ordinarily be very low probability, high impact events, so-called "Black Swans". This requires scenarios and thought experiments that are sometimes far outside the ordinary, and presented in very different ways. The goal is not to "predict" anything, because predictions under such conditions are a mug's game. Rather, the goal is to develop imagination and cognitive flexibility, so that when Black Swans do occur, one can respond adaptably and with agility.

And that's where this book comes in. Many different experts on infrastructure and urban systems have been asked for a short, relatively simple piece, be it poem, fiction short story, or technical essay. They are presented to you not as predictions, but as thought experiments: "what would you do if . . .?". The hope is that, as we learn to think more creatively even about the most mundane of urban systems and infrastructures, we will contribute to an efficiently resilient, desirable, equitable future – a scifi novel we would all like to live in. Take each piece as a scenario, enjoy it, and play with it. Bon voyage!

— The Cognitive Metainfrastructure —

Braden Allenby[*]

All infrastructure performs two functions. One is an explicit function that is usually quite apparent, and frequently involves a specific engineering domain. Municipal water systems are one example of a predominantly explicit infrastructure. The second function is as an enabler of other technologies and infrastructures. Electricity infrastructures, for example, perform the explicit functions of generating and distributing power, but they are also, and predominantly, enabling infrastructures for many other systems, such as information and communications technology (ICT) networks and devices. Each type of infrastructure mixes these functions in different degrees. Thus, for example, waste management infrastructures are primarily explicit; transportation infrastructure tends to be a mixture of both explicit and implicit; and ICT infrastructure, although it has a large and very physical explicit component, tends to be more heavily implicit. There are also interesting cases such as trade infrastructure, built not just on the physical backbone of transportation networks but also on the more virtual ICT networks of computerized global trading systems and information flows, as well as the financial infrastructures required for support. In general, when new infrastructures, such as electricity in the late 1800's, do appear, they are usually explicit and commercially important, and recognized as such by most people. Education and domain definition follow: thus, the rise of the electricity infrastructure also led to the definition and institutionalization of the field of electrical engineering.

But until now, different kinds of infrastructure have been conceptually if not physically bounded. Moreover, the traditional way of thinking about infrastructures, and developing the intellectual, institutional, and domain frameworks to design, operate, manage, and evolve them, has become so routine that it has obscured the rise of the first real metainfrastructure. A metainfrastructure is a complex adaptive system combining built, natural, and human elements that emerges from evolving networks of existing infrastructure systems,

[*]Arizona State University | brad.allenby@asu.edu

and yet is far more than simply the additive or synergistic behavior of coupled infrastructure systems.

We are now facing the first significant metainfrastructure in history, the cognitive metainfrastructure (CMI). It is not surprising that the CMI is emerging at the beginning of the Anthropocene, when a number of previously weakly coupled systems – human activity and the climate system, for example – are suddenly displaying signs of much stronger interaction. Constituent infrastructures and elements of the cognitive metainfrastructure are quite apparent, of course – 5G, AI/big data/analytics, social media, Internet of Things, increasingly powerful personal software and hardware, the Cloud, CGI and other media creation tools, social credit systems, weaponized narrative and modern disinformation warfare, and fierce policy arguments over privacy and identity, to name a few. But we have neither perceived this emergence of CMI from this increasingly complex and interactive set of earth systems, nor begun to grapple with what it means, much less begun to understand how we might respond to it.

The Oxford Dictionary defines cognition as "the mental action or process of acquiring knowledge and understanding through thought, experience, and the senses". As with most definitions of intelligence and cognition, this is a very anthropocentric definition – "mental" referring to human mind, and "thought," "experience," and "the senses" clearly anchoring cognition in the human sphere. But humans are only a part of the CMI constellation, and there is no necessary reason that the cognitive processes and outputs from the CMI will mirror human thinking and decision-making, with all of their frailties, heuristics, and emotional shortcuts. Indeed, given the many activities and infrastructures that integrate into the cognitive metainfrastructure, from sensor emplacement in billions of devices and infrastructure elements, to increasingly rapid and continuous machine-to-machine and chip-to-chip communication, to integrated techno-human and AI/big data/analytics enabled operation and management at all scales, there is no reason to expect this new metasystem to look like human cognition at all. Rather, the CMI operates at a level of complexity and information flow that humans can neither understand or perceive; people are low bandwidth cognitive mechanisms in a world where CMI operates at far higher bandwidth, and much faster speeds, than individuals can hope to access. What role human brains might play as CMI evolution accelerates is thus a very real, but very challenging, question.

Recognizing the emergence of this new metainfrastructure system from today's infrastructures and technologies has some practical implications. For example, in response to a bevy of complicated and interconnected policy challenges including security, privacy, speech rights, the role and definition of "truth," and the rise of vicious tribal politics, leaders in the United States and the European Union are proposing a number of policies which reach back to traditional frameworks, from anti-trust to the Constitutional idea of free speech. Knowing, however, that the situation involves the emergence of a conceptually new, more complex, metainfrastucture based on fundamental changes in the information environment provides a bracing reality check: just as one could not regulate Rockefeller's Standard Oil by using the regulations applicable to a local coal supplier, the attempts to use past understanding to manage the CMI are category mistakes doomed to failure. Worse, the comfortable retreat to past verities is a way of avoiding having to grapple with the real complexities attendant on the emergence of CMI.

Biography: Brad Allenby is President's Professor of Sustainable Engineering, and the Lincoln Professor of Engineering and Ethics, at the School of Sustainable Engineering and the Built Environment at Arizona State University.

— The Capsule Tunnel —

Amollo Ambole[+]

In 1899, the Western world stood on the brink of a machine revolution. Rudimentary tools were replaced by increasingly complex machinery: from needle and thread to sewing machines, from pens to typewriters, from hand-hoes to horse-driven ploughs. The innovators of these machines could hardly have predicted how much machines would transform human life. Their chief aim was to make daily activities easier. A hundred and twenty years later in 2019, machines are the fabric upon which the global society operates. Almost all industries, lifestyles, and activities are driven by machines and their accompanying technologies. And infrastructure has developed alongside the machine revolution; growing as the machines grew and dominating as the machines dominated. The roads are wider, airports smarter, train stations classier, and outer space has been explored more than once. The interaction between machines and infrastructure is a never-ending union and offers a world of new possibilities.

However, this transformation is not without a downside. The world is becoming more populated and people are becoming more individualistic. On a certain street in New York, there are twenty cars lined up on a traffic snarl up carrying twenty people - one per car. On another street in Johannesburg, there is a family of five leaving the house for work in five different cars only to be caught up in a traffic jam they helped create. Other than traffic congestion, advancement in motorised transportation and road infrastructure comes with other challenges like pollution and depletion of natural resources. A popular documentary, *Ultimate Vehicles*, showcases how the transport sector across land, sea, and air have advanced greatly in 2019. People and goods can move faster than ever. But as pointed out by the Intergovernmental Panel on Climate Change, pollution and environmental degradation are threatening all forms of life on earth.

Alongside technological advancement, communication has seen an upward trajectory. As the liquid transmitter was the most advanced form of communication in 1876, its innovators might not have imagined that people will have

[+]University of Nairobi | lambole@uonbi.ac.ke

handheld mobile phones in 2019. The Frankfurt Bauhaus would quickly become the first version of a rotary phone in 1925. 94 years later, communication devices are much smaller, immensely powerful, and more personal. Can these improved technologies in communication and transport merge at some point to become more sustainable?

It is now June 27, 2102 in the city of Nairobi, Kenya. The city's dwellers watch as star-like people enter and exit the city's skyline. There are no vehicles. What used to be roads filled with vehicles have been left for emergency services, recreation, and sports. The star-like people, a sea of commuters, are moving in small, personal self-driving carriers called *capsules*. Everyone has their own. Raheem, a young man who has just arrived in the city from Kampala, Uganda, looks up and wonders how this new form of transport works. He has seen them back home but not in such a large number.

The Capsule
Fifteen years back in 2087, the first capsules arrived in Kenya. They are made of biodegradable materials and powered by storable solar energy. The average capsule weighs eight kilograms, three of which are taken up by a recyclable battery and the rest by the lightweight materials utilized in the frame. On the outside, the surface or "skin" of the capsule is non-reflective, which causes minimum interruptions when it absorbs solar energy. Each capsule can transport a maximum of one adult and a child. The capsules detect the weight and number of occupants to determine the maximum speed; the speed is considerably lowered when there is a child occupant.

The design of the capsules allows them to stop mid-air and change trajectory. Inside a capsule, there is a screen that displays maps in a new 8D display. A user activates the same screen for all forms of communication, which allows the user to focus on tasks like watching shows, listening to music, talking on phone, dictating, or just sleeping as the capsule smoothly moves to the selected destination. Essentially, the capsule is ultra-modern transport meets ultra-modern communication; It can turn a drive to the park into a full-on virtual office.

How does one park a capsule? Seeing the capsules in the air is a fascinating thing for Raheem. However, he can't see any of them parked on the streets. As one lands metres away from him, Raheem approaches an occupant who takes him through the process of disassembling and carrying the capsule. Like a parachute, a capsule can be "folded" into a bag pack. The design of the capsule is such that it only switches off after the user has disembarked. The instruction is executed either by voice or at the touch of a button that allows different parts to fold, while others slide inwards to leave a portable capsule that can be carried around by the user. As the capsule folds, its "regeneration" is activated. This means recharging the battery of the capsule as well as repairing its self-healing skin.

What is the Capsule Tunnel?

For any form of infrastructure, amenities enable functionality. Proper operation of different forms of infrastructure requires key elements. The platform upon which the infrastructure works is critical to its operation. The platform is the basic mode of asset and service movement. A communication system enables interaction of different elements in the infrastructure environment. This interaction enhances efficiency and prevents eventualities like accidents. Lastly, the carriers facilitate end-to-end accomplishment of the infrastructure goals in an infrastructure system. The capsule tunnel is the platform within which the infrastructure operates.

The population of Nairobi city in the year 2102 is 19.5 million people. In 2019, the city could barely move around 4.5 million people due to its deteriorated public transport system. How then can the city manage over 10 million capsules in 2102? The answer is the . The capsule tunnel is a system of electromagnetic fields that control the movement of the capsules within the Kenyan geographical boundaries. It is defined as the roads for the capsules; invisible but efficient. The nature of the magnetic field operations of the capsule is called tagging. When a person enters a destination, the "tag" connects the capsule from the point of origin to the destination in a unique path. The tagging prevents the capsule from colliding with other capsules, thus facilitating fast and efficient movement. The generated routes are then connected through a special field that "pulls" the capsule from the point of origin to the destination through the generated path. Critically, the "tag" works in the same way as the internet did in 2015–it is owned by no one and regulated by no state, which shields it from legal and territorial conflicts. However, it is parallel to and independent of the internet.

The capsule tunnel also works with what is known as the "best route policy" that allows movement across any space as long as it is within the borders of a country into which the capsule was imported or made, using the shortest possible route. When an engineer working for *MuView*, a leading, open-source capsule manufacturer in the world is interviewed about how this works, he says:

7

"Geofencing is the trick. It is a technology from the late 1990s that allowed people to trigger an action within a certain geographical area. We do not intend for the capsules to be used for international travel, which is the essence of the one-time process of selecting the country within which the capsule will work. It cannot be changed, which means that if a person moves to a new country, he or she must obtain a new capsule that can operate in that country. The capsule tunnel is country-locked. It could be opened for international use in future but that would create socio-political challenges than we cannot deal with right now. Countries are isolated and the UN is weak. Meanwhile, multi-national companies are more powerful than ever. We simply cannot get governments to agree."

The concept of using electromagnetic fields in transportation works only on interlinked platforms. When the capsule tunnel was first developed, a comprehensive network of magnetic cables was laid in the ground across the country as was the case across many other countries. The magnetic cables create an electromagnetic effect across the country through which the capsules move. The concept of electromagnetism in transport therefore raises the question of the "pull" effect. Does the destination literally pull the capsule? No. The "pull" effect refers to the anchor effect where the magnetic attraction guides the capsule to the destination. The capsule is powered by its own energy.

Adverse Considerations
What could go wrong in the capsule tunnel? While explaining how the capsule tunnel communicates without causing accidents, Dr. Kimani, a tunnel sustainability expert from University of Nairobi, remarks:

"Take an example of a stormy day. Does it mean the capsules could fail? No. The capsules work with magnetic technology that is designed to survive such conditions. Could the uprooting of a cable cause failure? It should not unless a considerably large area of such cables is destroyed."

The capsule tunnel is designed to withstand adverse weather conditions and physical damage. Technology has evolved so much that magnetic commercialisation creates the possibility of weather resistant systems. As such, harsh weather is a challenge but not an inhibitor. The capsule tunnel does not work with lighting because it is an abstract path. There is no physical construction required for the tunnel to be operational except for the laying of the magnetic cables. Night conditions do not affect the transport system much because the traveller only needs to input the destination. He or she does not necessarily need to see how the capsule is moving because either way, it will stop at the destination.

The Capsule Tunnel Communication Infrastructure

2083 is the year that changed the communication landscape in transport. An innovation called the "tot" was introduced that allowed individuals traveling to similar destinations to locate each other using usernames that they set. This was an upgrade on a technology called "Bluetooth" that worked in the late 1990s and early 2000s. They would then interact and connect through peer-to-peer connections if they wished. The opt-in process was simplified to leverage the time available to the people. Inside the capsule, the communication framework is the same. Capsules can locate each other on the "tunnelwire" as a network of capsules that are powered within an area of six square kilometres.

As Raheem wanders about the city of Nairobi, he checks into a homestay that assigns him a capsule. He will use it during his stay in Nairobi, but it is programmed to operate within a limited radius around the city. He taps on the screen just as he had been taught and unlocks the capsule with facial recognition. He keys in a destination and sets off. Mid-air, he wonders what else he can do while in the capsule. He taps on a button which connects him to a voice. It is a woman talking so fast she sounds like a mad scientist. He does not know how to disconnect so he waits for the woman to stop talking. He starts sweating when she does not finish fast. Inside his capsule, the temperature drops considerably. He is so worried that he calls the hotel as he had been asked to do in case of an emergency. Another voice directs him to remain calm. He arrives at his destination safely. On his way back, he tries a few more options on the screen. One shows him red and blue lines, another one shows stars, another asks him to say certain words in case there is an emergency, while another asks him to initiate repairs. He does not select any of the options. When he gets back to the homestay, the owner explains to him what they all meant.

"The capsule is a self- monitoring, regulating, and repairing environment. The screen allows you to communicate with the capsule, the capsule tunnel, other capsules, and other service providers. It uses a technology called the Bosh that picks small packets of encrypted data called "tots"

and stores them on the surface of the capsule for them to be picked by relevant crawlers. For example, if you sent the message to another capsule, its crawler will pick it up and deliver it to the destination crawler. That process uses 12G network, so it is almost instant. The capsule automatically communicates with itself. If it senses an increase in sweating for the occupants, its internal temperature naturally and gradually drops while the reverse is true for when it gets cold. The red and blue lines you saw are other magnetic paths along the same route as you, which shows that other capsules are on the same trajectory. The "stars" are the other capsules."

Security

When it was first tested in Morocco in 2082, the capsule tunnel was a marvellous idea until a military officer in a capsule needed to respond to a shooting emergency. The capsule could not stop mid-air at the time and the officer had to go up to his destination before changing course and going back to the station. By the time he departed to respond to the shooting scene, the shooter had been neutralised, but only after killing twelve people.

"The 2082 incident got everyone thinking what security and emergency response would look like in the capsule tunnel. The challenges experienced and the testimony that followed led to the creation of a capsule cop."

The capsule cop is a stronger electromagnetic space enabled by more powerful capsules that are militarized and only sold to governments. Unlike normal capsules, it is a hybrid capsule that uses greater electric power to power its strong magnetism as well as improve its speed. It is operated only remotely with the occupant of the capsule being a "Robocop" - a robot that is designed for security and military work. The top speed of a capsule cop is thrice that of a normal capsule while its magnetic strength is four times better to give it higher accuracy and precision. In the capsule tunnel, a special path called the alpha path is created above all the other paths for the capsules to facilitate seamless movement for the capsule cop. The alpha path is called the "suave". Like the rest of the tunnel, the suave takes the shortest possible route.

Final Thoughts

As craving for efficiency and privacy increased, it became unsustainable to continue producing cars that pollute and increase traffic congestion. With a population of 14 billion in 2100, it would be insane to put car keys in the hands of even half the world's population. As such, the capsule tunnel can only get better. Further innovations will make the tunnel more efficient, more sustainable, and more accessible internationally. There are recommendations to make the capsules GPS sensitive although the GPS technology is long outdated. That would make them operate in different countries with simple changes in their settings. As concluded by Dr. Kimani:

> *"We see a future where everything sits in space. We see a future modelled on the capsule tunnel that would allow commercial offices in space with tunnel collaborations between employees. The capsule tunnel offers a model for communication as well. Small packets of information shared from one node to another that can be picked by individual capsules in work environments."*

The way of life in 2150 will be modelled on the capsule tunnel. There is a possibility that tunnels could be used not just for transport but to also improve housing, work, healthcare, and sports. That remains a dream to pursue.

Biography: *Dr. Amollo Ambole is a design researcher with a passion for sustainability issues. She has a background in product design and experience in development research. She uses her skills as a creative to facilitate co-design processes geared towards sustainability in African cities. From 2017 to mid-2019, she led an interdisciplinary research team in Kenya, Uganda, and South Africa to co-design better access to sustainable household energy in urban slums. Currently, she is a Principal Investigator in the SA-Africa-UK Trilateral Research Chair that is focused on gendered energy innovations. She hopes to continue empowering communities in Africa to co-design their own sustainable future.*

— Musings on Next Generation Infrastructure: The Art of the Sweet Spot —

Adjo Amekudzi-Kennedy[*]

In our quest for next generation infrastructure that is at once functional, resilient to climate and other disasters, sustainable, and with other characteristics we value, it would be wise for us to intentionally identify and innovate our adaptive methods as mitigating ones, simultaneously. The sweet spot of adaptation and mitigation is one that is more sustainable, because it helps us adapt to the effects of the changing climate and other hazards while reducing the causes of climate change and related hazards. Otherwise, we might adapt ourselves into oblivion. Likewise, considering opportunity, we must intentionally find opportunities to innovate that enable us to simultaneously adapt to while mitigating the effects of climate change. This sweet spot of adaptation-mitigation is a space for creating sustainable infrastructure; otherwise our infrastructure may need to be adapted over and over and over again ... into oblivion.

Adapting without intentionally identifying and incorporating mitigation opportunities is inefficient, a missed opportunity, and the path of weak sustainability. Adapting while innovating infrastructure facilities and systems to contribute increasingly to mitigation is a path to strong sustainability, affecting not only survival but incrementally developing built, natural, social, economic and other capital resilience to existential threats.

Our infrastructure planning approaches have long been predicated on forecasting, more recently backcasting and scenario planning. While helpful, we have missed the mark many times with our forecasts, especially as deep uncertainty becomes more common and stationarity increasingly insufficient for decision-making. Backcasting and scenario planning have improved our understanding of how to shape an attractive robust future. We live in an increasingly dynamic world however, and must plan dynamically. But not dynamically abandon. Dynamic adaptive planning may help us save the day if we seize the opportunity to adapt while trying to address the factors that led us to anthropogenic climate change in the first place.

[†]Georgia Institute of Technology | adjo.amekudzi@ce.gatech.edu

The proliferation of new technologies is wonderful in urging us toward a future of more and exciting choices, higher quality of life, economic competitiveness, social cohesiveness. It can also be dangerous when predicated on assumptions that only technologies can provide lasting solutions to our most pressing needs. Without engaging human behavior and intellectual capital, our definitions of "smart cities" may never be complete. Our cities will become smarter as we learn to develop and harness new technologies and data, develop and apply our understanding of human behavior and technology interactions, value basic dignity in all humans, develop community resilience and draw our communities toward the outcomes they value.

Many will agree that infrastructure is an enabler. An enabler of mobility, solid waste management, communication, access to clean water, access to clean air, access to education. Infrastructure inadequacy or lack leads to unfulfilled needs, desires – propagating into quality of life deficiencies, economic deficiencies, environmental deficiencies, communication deficiencies... For infrastructure to be effective and efficient, it must be available at the same levels of quantity and quality to all — in the long run. Infrastructure equity will therefore continue to remain desirable in regions of progress and less desirable in regions of little progress. And communities with higher levels of infrastructure quality and quantity can expect to create more wealth – multidimensional wealth that elevates human quality of life and advances the economy with no sacrifices to the natural environment. Equitable infrastructure has, in regions of innovation and development, proven to support more resilient communities – from the standpoints of short term (resistance), long-term (recovery) and adaptive resilience.

Infrastructure systems in the future ought to be as adaptable as they are sustainability-oriented, that is augmenting quality of life in economically efficient and environmentally regenerative ways. Most of our infrastructure sits in a commons. Leaving parts of the commons behind will leave gaps in a community's fabric of resilience.

Stewards of infrastructure and leaders of tomorrow who find ways to crowd source the brainpower of entire communities through information and communication technologies (ICT) will have a competitive edge. For who will want to dream with one mind when he or she can dream with hundreds, thousands and millions of minds?

Biography: Adjo Amekudzi-Kennedy, Ph.D. is a Professor in the School of Civil & Environmental Engineering at Georgia Institute of Technology. Her research, teaching and professional interests are in Infrastructure Asset Management, Transportation Planning, Sustainability Modeling and Policy Development, and Engineering Leadership Development. Amekudzi-Kennedy is co-author of the textbook: Systems Engineering with Economics, Probability, and Statistics (2012), and has authored and co-authored over 100 refereed journal articles, technical reports and other publications contributing predominantly to advancing infrastructure asset management and sustainability thinking. She is the Founding Chair of the American Society of Civil Engineers' (ASCE) Committee on Sustainability and the Environment, Founding Co-Director of the Global Engineering Leadership Minor at Georgia Tech, and has served as a member of the Board on Infrastructure and the Constructed Environment (US National Research Council, 2009-2018), and chaired the 2009, 2014 and 2019 Advisory Boards for the ASCE Infrastructure Report Card. In 2018, Amekudzi-Kennedy was elected as a fellow of the ASCE, and received a Georgia Power Professor of Excellence Award from the College of Engineering, and a Class of 1940 Course Survey Effectiveness Award from the Institute. She was elected to the National Academy of Construction in 2019.

— My Computer is in Love —

Clio Andris[†]

My computer has been acting strangely over the past few weeks. He's in his own world, daydreaming and languorous. He opens up programs to create artwork and prints them out all over the house. I came home yesterday to 200 drawings of acoustic guitars and lilies! He puts on romantic movies when I need him to be answering messages.

Last week he downloaded 17,000 different recipes for beef bourguignon. I don't eat beef! He doesn't even eat at all! I saw that he ordered candlesticks and a tablecloth to be sent to the house.

He booked a flight and accommodations while I wasn't watching. He now has his own reservations and I can't see where! But my statement says he booked them and he has all of my banking information and passwords.

We are seeing this kind of thing happen from time to time. There was the lamppost outside who got depressed and only flickered in the mornings. And the submarine who found religion and aborted a bombing mission. The street camera that went on strike until he was given a mini jacket to wear. Then there was the treadmill that had to go to sensitivity training. The couch ottoman that followed its owner around wanting feet attention. Then the anxious train intercom that was too self-conscious to make an announcement for a whole week, and passengers had to make him feel better. And the pizza delivery drone, to spite the pizza-maker, that dumped pepperoni pizza deliveries into the river! There were dozens of pizzas in the river!

And before this recent string of incidents? Very little. I found one a historical account of a textile factory in the 1800s: After a new chimney sweep was hired, one of the looms apparently started weaving blankets with hearts all over them, and no one could figure out how or why. After the chimney sweep moved on, the loom returned to weaving just stripes. A loom! It had no central CPU, only loom parts.

[†]Georgia Institute of Technology | clio@gatech.edu

I think my computer is in love. Again. This has happened once before, with the toaster. He tried to serenade it and organize romantic dates. He wrote it poetic SQL queries, and took a lot of pictures of it. A lot of pictures. I had seven million pictures of the toaster at various angles and times of day. There was no room for any of my files. But, ultimately, the toaster did not respond and the computer went to the edge of the bathtub to plunge in and I said, no, wait, you're not waterproof! He came down and it was a few days until it booted back up and regained its speed. But I try not to bring the computer into the kitchen to work at the kitchen table, out of sensitivity. And, the toaster seems fine; it still toasts bread.

Now I think he has fallen back in love. But with what or whom? I don't think it is one of the plants. Perhaps the vacuum. It could be the neighbor's dog. The thermostat. The iron. One of my sweaters. The toaster, again? I looked through his older memory to see what he has been thinking about, but saw that he had learned to erase this. Good. Good for him, I thought. My computer deserves his privacy...

On his desktop he generated out a list of new goals for his look. He wants a haircut and new clothes. He doesn't wear clothes (but then he learned about the camera's little jacket) and he has no hair. Still, I told him I would take him to the barber shop, in hopes that maybe we could get back to work afterwards. I didn't make an appointment—what would I say? Please cut my computer's hair?

We arrived at the barbershop. "My computer would like a haircut." I announced to the barber regally and pointed to the computer ceremoniously, hoping the barber would understand that I was playing along with the computer. The barber looked me and then the machine to decide which of us was ill. "He told me this today," I continued and winked. The barber further scrunched his nose.

I took the barber aside. "He has one of those things going on, like with that the train intercom, or that dreadful treadmill," I whispered. The barber nodded, bringing up the story about the submarine. "Exactly," I replied.

"You're on the list!" I told my computer. He beeped and spun around some numbers. We sat in the waiting chairs for a bit while the he pulled up the world's supply of leather jackets. The barber returned ten minutes later and kindly escorted him to the chair, which he raised all the way up.

The computer pulled up some images of movie stars and the barber nodded and started snipping in the air and humming. The barber brushed him off and said "No charge. Thanks for doing my taxes." My computer looked in the mirror and snapped one proud shot.

I'm happy for my computer, but it is getting expensive and it is hard to get work done. Usually, we would be working together and he would be engrossed in his normal hobbies, work and finding cute pictures of cats.

Instead, he just bought a leather jacket and a gym membership. He's not listening. "Now we need to work! It's a work day!"

He giggled. He never giggles. "Stop giggling!" He's dancing. He's writing songs. I think he's in love with the toaster again.

Biography: *Clio Andris is an assistant professor at the Georgia Institute of Technology where she directs the Friendly Cities Lab.*

— Here we go again —

Weslynne Ashton[†]

For more than fifty years, we have had no real home.
They still call us climate refugees.

I remember my island, my first home,
How the rains made every patch of land bloom in a thousand shades of brilliant
green,
How the sky touched the sea and you could lose yourself in all that blue
Water,
Everywhere,
Pelting from the sky, the sea, swirling, rising, covering
Everything,
Gone. Our house, our neighborhood, only the stench of rot, wood, carcasses.
We went back to salvage what we could
Before they declared it
Uninhabitable,
Too expensive to rebuild.

My children, my grandchildren, will never know cool, island breezes, dancing
through their hair.
Now tropical islands are nothing but
Too hot, too small and too hazard prone for poor people to live…
Only the rich can afford to visit the micro-islets,
What were once the tops of island mountain ranges.
They fly in on hyper-copters to vacation inside floating, air-conditioned bub-
ble zones,
The hotels serviced only by robots,
Food flown in from across the globe by drones.
They enjoy diving expeditions to see what's left of volcano craters and tropical

[†] Illinois Institute of Technology | washton@iit.edu

ocean life.

We were evacuated,
Relocated
To a Refuge City in a foreign country.
At first, they said it was temporary.
But you stop counting years after your children have children.
We were happy that we found somewhere to go,
That somebody was willing to take us in,
But it's hard to feel joy amidst
All this gray and brown.

Our new home was nothing like the refugee tent camps of the early twenty-first century.
They are self-enclosed cities,
Self-sufficient, but separate from the rest of the host country.
Our gray waste-plastic, 3D printed concrete houses, are sturdy, but sterile,
Strong enough to withstand a category 10 cyclone they say.
Each house generating its own microturbine power, feeding into the City's microgrid
Each circulating its own water, for drinking, for cooling, for washing.

But the land around us is all gray and brown, too wasted for life to grow.
The trees are gone, the bees are gone.
So they moved the factory farms inside buildings.
They gave us each our own pod
To grow some of our own food, vegetables and nu-meat,
To build community through agriculture,
To talk and learn from others who didn't have the money, education or connections
To migrate to the Great Cities of the East.

We who are without money, without a home, without a vote
What voice, what choice do we have when they say what we must do?
We are subject to the whims of our hosts, they share dribbles of their technology and
infrastructure
But keep us

Dependent and
Separated
From their people.

Now we see
These seas are rising,
These aquifers are drying up and must be abandoned,
Despite all the robots and engineering efforts to stop the salt water from intruding.
This land, the heat, the drought will force us to move
Again.

Except now, it won't just affect us, but them too.
If only, they had thought in 2050 to ask us to share
Our traditional knowledge,
Our culture,
Our science and engineering,
Instead of only seeing us poor refugees, to be pitied and pushed aside,
Perhaps we would not be here
Again.
But the wisdom we once had
Is lost
Now it is too late.

We can't help you now.

—-

As we build new physical infrastructures, which could last into the 22nd century, we need to consider how people will participate in these systems, and be given voice to control how their lives take place within them. In particular, we might consider what can be learned from those who are displaced from their homes, because of climate, economic, political or cultural change, and how their knowledge and experiences can be woven into creating more resilient infrastructures. We might consider how to build soft infrastructures into our technological and infrastructural advances, so that individuals gain the capacity and opportunity to participate meaningfully in their design and operation. We need to accommodate diverse voices, particularly those most vulnerable to climate disruptions. We might consider the complex interactions of multiple social, ecological and technological systems, across individual, organizational, regional and global levels. We must design infrastructures capable of adapting to changing biophysical conditions and sociocultural needs, which we can only speculate upon today.

Biography: Weslynne Ashton is an Associate Professor of Environmental Management and Sustainability at the Illinois Institute of Technology Stuart School of Business. Dr. Ashton's research focuses on industrial ecology and the circular economy, or how to optimize energy, water and material resource flows as well as human benefits in socio-ecological systems. Her work also examines developing entrepreneurial solutions to social and environmental challenges and the adoption of socially and environmentally responsible strategies in business, particularly small and medium enterprises (SMEs).

— The infraordinary importance of infrastructure —

Aristide Athanassiadis[*]

Infrastructures are boring. Who cares about pipes, concrete and weird liquids when we are facing serious societal and environmental challenges? At least that is what I thought about infrastructure by the end of my Architectural Civil Engineering degree almost ten years ago. I felt that designing structures, superstructures and infrastructures was a thing of the past and of less interest. I thought that engineers were only interested in pouring concrete and working on excel sheets. I needed to focus on more pressing issues that directly have an impact on my everyday life.

That is why I decided to shift my focus to study cities. I studied them from a historical point of view, an economic one, a social one, well plenty of different facets that compose the complex systems that our cities have become. Discovering more and more the environmental impact of cities, I then decided to specifically focus at the environmental side of cities from a systemic point of view. In other words, study all the flows of resources and waste entering and exiting cities, as well as the actors governing them. Some people call this an urban metabolism study, or an urban metabolism approach. As much as I wanted to get rid of my engineering "background", these studies entailed to collect data and analyse it in order to better understand the resource requirements and the pollution/waste generation of cities AND optimise them.

Here I am stuck with numbers and optimisations again. A different topic, but the same techniques. But the more I continue to dig deeper, the more I discover that behind the data and figures are physical and material artefacts. In the "developed" world, we often forget all the intricate infrastructure that enables us having bright light at our homes by the simple flip of a switch, having drinkable water by turning on the tap, getting rid of bacteria-filled human defections by flushing, or simply ordering on internet any kind of product and receiving it by the following day. The amounts of copper, cement, sand, gravel,

---

[*]Université Libre de Bruxelles | arisatha@ulb.ac.be

steel, energy, water and greenhouse gas emissions that were needed to enable our modern way of living is mind boggling.

Not only that, you also have to think at the spatial imprint of these infrastructures. To receive water, you not only need pipes, pumps and reservoirs all over your city you also need some upstream infrastructure. You need dams and water treatment plants that can be tens or hundreds of kilometres away. For energy and products, it is even worse. Just imagine how many steps it takes to make electricity in your home town regardless of the initial source of energy. We either import fossil-fuels from other parts of the world to be burned in energy plants (that we need to build with imported materials), or import materials to produce local renewable energy.

Believe me, the deeper you dig, the messier it gets.

But that's not it! This is just a dry engineer's perspective only looking at materials and numbers. What I came to understand is that infrastructure and numbers are also political beings. In fact, infrastructures not only embody materials but also political choices, negotiations, struggles and conflicts. Some of these choices are still manifest in a number of European cities but you have to know where to look. Paris and Brussels are prime examples of the 19th century hygienist urbanism. An urbanism which aimed to eradicate misery and diseases (but also social unrest), by developing huge infrastructure works. In many cases however, such infrastructure works were synonymous with displacing poor population for tree filled boulevards, eliminating noisy and smelly craftsmen (crucial for what we would call urban manufacturing and circular economy), and enabling military access to the city centre to repress any rebellion and civil wars. In the case of Brussels, the Senne river was completely covered and transformed as a sewer, completely disconnecting the city from its environment and radically altering material flow cycles.

The social, political and urbanistic consequences of the choices taken almost 200 years ago, can still be seen today. In many cities, canals, train tracks, highways, have created socio-spatial fractures, isolating segments of population, decreasing social mobility and altering the urban development of cities. This also implies that depending on the design and implementation of infrastructure different segments of population will have access to certain services and flows. Green spaces, quality of water, waste removal can be experienced very differently in different neighbourhoods within one city. In addition, while a number of infrastructures well traditionally owned, built by and maintained by the public sector, they are now more and more operated by the private sector furthering the access, cost and equity challenges.

The long-lasting (and sometimes undesired) effects of poorly designed infrastructures can therefore lead to a "lock-in" effect, forcing us to maintain obsolete infrastructure because they were "just too expensive" to build. In the advent of material criticality, circular economy and such concepts have made incinerators outdated and require new solutions fast. In fact, this points out how difficult it is to predict future uses and to design infrastructures that are adaptive yet efficient.

So, I guess this short personal reflection became an unwilling ode to infrastructures. Not necessarily to infrastructure planning and management but more to acknowledge their central and pivotal role towards a functioning society. As Perec would put it, infrastructures are the opposite of extraordinary, they are infraordinary. They sit just under our nose and surround us for our every action and except from the few cases where infrastructure failures are brought in the news, sustainability and unsustainability of infrastructures are rarely put forward. They are just performing silently their function to fuel our societies.

All the above mentioned brings us to probably the biggest challenge we are and will face for the next decades: the global population is urbanising fast and most of these future urbanites will live in cities that do not exist today. At a moment where resources are scarce and pollution is abundantly present in various forms, the future of our societies might rely on the future of our infrastructures. Infrastructures will need to be co-designed by numerous stakeholders and disciplines which are clearly not confined in the engineering faculties. They will need to negotiate options between centralised vs. decentralised, private vs. public, low-tech vs. high-tech, fixed vs. adaptive, etc. and lead us to different futures.

All things considered; infrastructures might not be as boring as I thought.

Biography: Aristide is Chair of Circular Economy and Urban Metabolism at the Université Libre de Bruxelles. During the last years, he has collaborated and worked with/for several universities, research centres, environmental administrations, NGOs, youth organisations and consultancy firms on a great variety of projects. Finally, Aristide co-created the non-profit organisation and open-source urban metabolism initiative Metabolism of Cities that aims to bring together people, data and publications in one central place.

— The Circular City —

Callie W. Babbitt & Nora C. Babbitt[+]

The city of the future – like the image shown here – will be made up of things we now consider to be 'waste.' The growth of urban populations will no longer be fueled by scarce resources that are extracted, used, and thrown away along a wasteful linear process. Instead, urban mining will provide a closed-loop source for obtaining vital natural and technical materials. Infrastructure will be designed for long and adaptable life. And when our buildings, vehicles, food, clothing, and computers have been used to their full potential, the energy, water, and critical materials they contain will be reprocessed and repurposed to meet societal needs. By reimagining 'wastes' as resources, we will not only sustain the environment we depend on to flourish, but also the health and vitality of our circular city.

[+]Rochester Institute of Technology | cwbgis@rit.edu

Biography: *Dr. Callie Babbitt is an Associate Professor in the Golisano Institute for Sustainability at Rochester Institute of Technology, where she conducts research to proactively quantify and minimize environmental impacts of emerging technologies. Callie's research group creates new methods and models in the field of industrial ecology that are inspired by the study of ecological systems in nature. They apply these models to study sustainability challenges and solutions for food waste management, consumer electronics, lithium-ion batteries, electric vehicles, and nanomaterials.*

Nora Babbitt is a student in the Brighton Central School District in Rochester, New York. She aims to use photography to raise awareness of environmental and animal welfare issues. She assists on research and outreach projects related to sustainability in e-waste and food waste management systems.

— In the Year 2121 —

Lawrence C. Bank[*]

Sept 2019

So, what will infrastructure look like and what will our infrastructure problems be in the year 2121? This is what I think.

No doubt climate change (or adaptation) and population growth (or degrowth) and resource availability (or scarcity) will have had, and will continue to have, their impact by 2121. As will the numerous and ongoing wars and geopolitical skirmishes.

But a hundred years is not that long a time. Those of my age (the baby boomers) can remember the 1960s and also the world after the 2nd World War. My grandmother was born in 1901. My children were born at the end of 1900's. They may well be alive in 2100 and remember the world of today. It will not be that different. On the other hand, if you lived through the 19th century, that was not so. That hundred years was a watershed. I don't think the 21st Century will be one of such magnitude. Maybe the 22nd. Probably the 23rd.

Man will continue to battle nature, and much of the natural world as we know it will have disappeared. But, I believe, there will still be some lions and tigers and elephants, many insects and birds, and many trees and plants and lakes and rivers and oceans. They will all be different but they will still be. And there will still be highway bridges and tunnels and sewers and water pipes and electricity cables to contend with. Resources will be different but limestone and water (at the very least saltwater) and iron ore will still be here to construct with. Agriculture and food production will be different, but we will still be able to produce enough food for the world's population. Power sources will be

---

[*]Georgia Institute of Technology | lbank3@gatech.edu

different, mostly renewable, but readily available – especially in the developing world which has been largely deprived of power up till now.

So, what will be that different? The difference, I believe, will be in distribution and storage. This may sound rather pedestrian. But these, I feel, are the key issues and dangers for the near future (the next 100 years) that now are coming into focus. It will not be if we can produce enough (whatever that may be) for the world's population but whether we can deliver enough to the places and people with the needs and wants. With 80% of the world's population living in cities or large metropolitan areas, the means of delivery and storage will become paramount. Especially since the basic urban and rural infrastructure will be very much the same as it is now – roads, highways, rail lines, air lanes, sewers, water lines, buildings, farms, and dwellings.

So, we will need to have the capacity to store water, power, food and medicine at the point of use and service for reasonably long periods – one to two weeks, perhaps up to one month. And we will need an infrastructure system that can deliver a week to a month's supply of these basic resources in an efficient and reliable fashion.

So, what will this look like? We can already see. Large secure, warehouses located close to major metropolitan areas. Fleets of delivery vehicles for land, air or other means. But delivery in bulk to local areas on a scheduled and reasonable basis. Not the delivery of the unsustainable single "gig economy" item from Amazon or The Amazon. But container-sized delivery to a local distribution areas for redistribution by foot.

So, if so, then we will need significantly more local storage on a single residential building or group of buildings or single- or multi-family dwellings (an eco-district model of sorts). But land in urban areas will be expensive and scarce. So, by 2121, I believe, we will see much more of the current (and excessive) urban living area space converted to storage space. This we already see in the ubiquitous mini-storage facilities around urban areas in developed countries. But instead of storing unused and sentimental items, we will store our vital, replenishable resources. Not for the day of reckoning but for everyday use.

So, the future I see in one hundred years is part on-grid and part off-grid. Densely populated urban micro-districts served by industrial areas with large open spaces for leisure and nature. I guess that may make me an optimist.

---

*with apologies to "In the Year 2525" Zager and Evans, Released Jan 1, 1969.

Biography: *Lawrence (Larry) Bank is a Research Engineer at the Georgia Institute of Technology and is a consultant at LCB Consultants LLC in New York City. Dr. Bank served as a Program Director at the US National Science Foundation (NSF) in the Materials Engineering and Processing (CMMI/MEP) program from 2014 to 2015, and as the Program Director for Structural Materials and Mechanics (CMMI/SMM) program from 2008 to 2010. From 2010 to 2013 he was the Associate Provost and then Vice President for Research and Sponsored Programs at The City College of New York. From 1975 to 1976, Dr. Bank studied in the School of Civil Engineering and then the School of Architecture and Urban Planning at the University of Cape Town (UCT) in South Africa. Dr. Bank received his BSc degree from the Technion in Haifa in 1980, and his MS and PhD degrees from Columbia University in New York City in 1982 and 1985, respectively. He has previously been employed as a Structural Engineer with Leslie Robertson and Associates (LERA) in New York City, and as a professor at Rensselaer Polytechnic Institute (RPI), The Catholic University of America, the University of Wisconsin-Madison, and the City College of New York (CCNY). He is the author of the textbook "Composites for Construction: Structural Design with FRP Materials," (Wiley, 2006), over 200 technical publications, and 11 patents and invention disclosures. He is a Distinguished Member of the American Society of Civil Engineers (ASCE), a Fellow of the American Concrete Institute (ACI) and a Fellow (and Past President) of the International Institute for FRP in Construction (IIFC).*

— The Disappearing City —

Michael Batty[†]

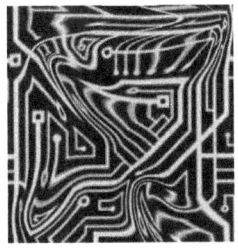

An Essay from the Future: My grandfather was born just before the end of the Second World War in 1945, my father in 1976 and I just as the world entered its era of no growth, or austerity as it was called back then in 2010. What was remarkable when I was growing up, was the feeling that all of us were not very rich compared to those who we met from other countries in the developed world, including much of Asia Pacific. We were always under pressure financially but, everywhere around us, new building was taking place. How was it, we asked, could so much money be pouring into office blocks and high rental housing when transportation and affordable housing was so under-resourced? Wages seemed to be continually eroding, taxes stealthily increasing and interest rates on capital were at historic lows. By the time we entered the 2020s, it was widely acknowledged that the wealth, accumulated over the previous half century by an aging population, was slowly but surely ebbing away, captured by smaller and smaller enclaves of the super-rich.

Debt seemed to be the name of the game – if you could borrow, you did so, taking on commitments that you never worried very much about. You would never save; only the elderly tried to save but with interest rates much lower than inflation, it seemed to be a complete waste of time. In fact, the 2020s became the decade of negative interest rates and resulted in you paying the bank to protect what monies you had accumulated. Set against a background of worsening pension plans and with the young having little prospect of owning their own homes, became an ever greater challenge. Unemployment everywhere on earth which appeared to have been rising in the late 20th century, had in fact never fallen and it was considerably greater than the official statistics revealed. Unofficially in developed nations, it hit 20 percent at the time when the UK left the European Community in the early 2020s and it reached this level a little

[†]University College London | m.batty@ucl.ac.uk

later when President Trump changed the law to enable his re-election for a third time. There were new jobs of course in AI and other high tech but automation itself was slowing their increase, and an increasing proportion of the potential labor force was underemployed and somewhat impoverished.

Into this milieu in 2020 came the Coronavirus, a widely unanticipated pandemic that led to many deaths and a dramatic drop in GDP. Many hailed this event as a 'warning from on high', sometimes taking on religious connotations. However in the early 2020s, a recovery of sorts began after various vaccines were discovered but as the world limped on, the old trends quickly reinforced themselves. The great unravelling actually took place in the third decade of the 21st century and hastened a perfect storm of populist politics, decreasing real incomes, and worsening and ever more extreme polarization. Fewer and fewer people were required to manage industry and commerce which were being automated at a rate faster than anything seen so far during the previous 250 years. Warren Bennis's story, probably apocryphal, about the automated factory seemed to be coming true: "The factory of the future will have only two employees, a man and a dog," said Bennis. "The man will be there to feed the dog. The dog will be there to keep the man from touching the machines." To an extent, our most educated populations still managed to acquire reasonable wealth by innovating and advancing new forms of technology. But increasing wealth was largely generated by using money to make money. This ultimately meant great wealth was being accumulated by an increasingly small fraction of the population whose expertise lay in the deftness of their ability to figure out how to become rich quickly.

Information technologies of course continued their relentless march with more and more automated systems coming to replace manual technologies. These began to lower productivity as entire populations struggled with layer upon layer of security to keep out users unwilling to pay and to keep in users who were already buying such services. During this time, the first autonomous vehicles emerged, and although forced by the very low price of oil which underpinned the move to electric vehicles in the mid-2020s, problems of mixing autonomy with the complexity of urban environments that had been created sometimes over hundreds of years, dramatically constrained their adoption. Many new regulations emerged, most to reduce accidents due to mixing with traditional vehicles and pedestrians, and these restrictions made autonomous vehicles for most users rather unappealing. By the end of the 2030s, the first robots – personal assistants – walking on the streets appeared. These too were highly constrained as much by their owners owing to the fact that they were expensive and vulnerable to violence from an increasingly polarized population who considered these machines fair game for exacting revenge on the elite that created them. In fact, urban crimes against humans continued to fall while crimes against robots and related automated machines soared. In short, automation appeared to be stalling. In the previous 100 years, automation had become highly routinized and simple to manage, but in the new digital world where new software was being created all the time and disrupting old, there was a slow but sure return to traditional, manual methods.

By the 2040s, there were very strong moves against multinational companies whose activities for years had tended to bypass national governments' taxation and the likes of Facebook and other digital platforms were hit with massive bills. Their profits quickly eroded and many of them went into a tailspin from which they never recovered. This was to a large extent the death-knell of globalization. It had been heralded for many years by strong moves to localism and by the pandemic which quite clearly had been dramatically accelerated by a world which was crisscrossed by travel, information transmission and global supply chains. Combined with slowing transportation in cities, unemployment rates growing upwards of 30%, and an increasingly small proportion of the superrich owning most of the assets and resources of an increasingly 'developed' world, saw many national governments increasingly focusing on local matters. The world, it seemed, was drawing inwards.

By the middle of the century, climate change due to human impacts had become a widely accepted view. Although there had been strong movements towards a low carbon future for over 50 years, for many nations, including China, this change in pace had never been fast enough. Although national governments were beginning to seriously counter rises in sea level and urban temperatures with multiple initiatives, these tended to be too little, too late. With well over 70 percent of the population living in cities of various sizes by 2060, and with well over half of this population living within 50 kilometers of the coast, city governments everywhere were having to redouble their efforts to combat climate change. Combined with the economic crisis, this reduced the world's collective wealth and resources even more. The prospect of a future based on a combination of 'Mad Max' and '1984" looked increasingly likely. Certainly no Brave New World was emerging.

In the early 2000s, the notion that successive cycles of technological innovation were crowding in on each other, getting ever shorter, led to the forecast that society was headed towards a singularity of enormous proportions. Some argued this would lead to entire societal breakdown while others such as

Google's Futurist Ray Kurzweil hailed this as a great opportunity for populations to achieve immortality due to rapid advances in medical technology. He predicted that this would have taken place by 2045 but the idea of a population based on half-man-half-robot, living forever, had always appeared far-fetched. In fact, it was our inability to jump out of the economic tailspin begun in 2010 that was the prime obstacle to rapid deployment of new medical technologies and the consequent inability of populations and governments to mobilize themselves to produce the required resources. As for many aspects of automation throughout the 21st century, those who both controlled and got the greatest advantage from them were those who formed part of an ever smaller elite. In an increasingly polarized world, there were few policies addressed to reduce polarization and segregation and most were simply intent on the alleviation of the poverty that had begun to beset many parts of the developed urban world.

Of course, it was the China crash in the early 2050's that really convinced us that the world was headed towards a kind of oblivion. The Coronavirus crisis in 2020 had been in the vanguard of this but the world did recover quite quickly although this was a far cry from the optimism of the previous century and earlier industrial revolutions. There had been crashes in the Chinese economy before but these had always been propped up by the central state. This consisted in increased debt by printing money, by the construction of more and more housing for speculation, and by the gradual push of the Chinese economy westward into central Asia in support of the long standing but increasingly failed policy of 'Belt and Road'. By the middle of the century, the bubble that had lasted more than 60 years had really reached its limits. When it burst, the world's various economies, already struggling with climate change and social polarization, were thrown into a dramatic crisis. This began the great transition to a very different world, one which ultimately would eschew technological change and begin its retreat to a simpler, more predictable, less frenetic world – a world where new and powerful regulations on how technologies could be used and owned were rapidly being put in place. The China crisis resulted in a rapid and massive decline in world trade, and many economies began to deploy more traditional technologies with respect, for example, to transportation. There was a sense of all this in the Coronavirus crisis in the early 2020s, but this was much more of a rapid world-wide shock where from the 2030s onward, technological change had seemed ever more problematic. The digital world of course still remained the key technology, but the idea that big cities were better than small – an idea which had been under scrutiny anyway since the 2030s – was rapidly being dispelled as the path to the future.

The most-deep seated crises took place in the biggest cities and within cities of perpetual economic turmoil. Housing markets began to collapse as did Fintech, while populations began to vote with their feet, moving to smaller towns. The frantic nature of the modern economy began to diffuse and dissipate and, slowly but surely, a sense of balance began to be restored to those economies that had best avoided the world's economic tailspin. Many of the prestige projects, such as those involving the exploration of outer space, came to an end – with little economic consequences because they had only ever been vanity projects for the super-rich or for nation-states. The move back to the countryside began earnestly in the 2060s, and inward-looking ideas about self-sufficiency dominated our thinking of the future. Increasingly the trend of working from home came to dominate. Of course the infrastructure that had come to characterize our cities during the first half of the century – poor quality construction of high blocks, ghost cities all over the most recently developed world in places like China, and the high speed train fetish that increased polarization largely subsidized by governments albeit indirectly never making any kind of profit, came to an end. All these features of our cities represented the basis on which this new move to a smaller, greener, more self-contained world began. In comparison with the gleaming towers of the world's biggest central business district in cities like Shanghai, New York and Singapore, most urban development by the 2060s was more shambolic and low quality, riddled with congestion, and quite simply unaffordable for most urban populations. The withdrawal to a smaller, simpler world proceeded apace; and as we headed towards the end of the century, a slower urban world emerged.

Let me finish with a story from the industrial era, over 150 years ago. In 1909, the British novelist E. M. Forster wrote a little story called "The Machine Stops". In it he painted a picture of a world in which physical transport was largely abhorred and men and women had retreated to their homes underground, communicating digitally through 'The Machine'. This is a picture of world similar to the one we are heading towards but unlike Forster's story where the machine does literally stop and all civilization descends into chaos, this is only one prospect as much because we cannot determine the future beyond the time we are living in. With a world population now pretty much in the steady state of zero population growth, we still have the ability to decide how we might reach a better world. This is likely to be much less focused on the greed associated with naked capitalism and the social polarization which has dominated the century so far, and much more around an earlier vision of cities working in harmony and exploiting the best of our technologies in the most equitable way. As we are all no longer heading to immortality, our speculation about what the world of the 22nd century will be like, is a bridge too far but doubtless, this future will be very different from our current world of the late 21st century.

Biography: Michael Batty is Bartlett Professor of Planning at University College London where he is Chair of the Centre for Advanced Spatial Analysis (CASA). He has worked on computer models of cities and their visualisation since the 1970s and has published several books, the most recent being **Inventing Future Cities** *(MIT Press, 2018).*

— How will city dwellers get around in 2100? Personas using future mobility services —

Lynette Cheah[†]

In user-centered design processes, personas are sometimes developed to better understand how potential users will make use of new products or services. Personas are fictional characters projected into future settings and situations. They are created to represent people with different backgrounds. By identifying and characterizing representative users, designers and planners can better address their diverse needs and concerns.

In this chapter, we get to know four different personas living in future Asian cities. Specifically, we explore how these individuals interact with urban transport infrastructure and achieve their mobility needs. It is not possible to predict how city residents will access needed goods and services in the year 2100, but we can imagine their urban lifestyles through these personalities.

We focus on archetypes of city dwellers in Asian cities, which is expected to host a significant portion of the world's population by the end of the century. Together with visions of the state of transport infrastructure in 2100, we can empathize with their needs and concerns, which allows us to better understand what their daily activities and travel behavior might be like.

City dwellers today and tomorrow, like June, Sumit, Chao, Sam, and Sabreena,[*] will adapt their activities and travel behavior according to the prevailing technologies and available services. The snapshots of their life events are captured via personas as narratives of a plausible (or imaginative) future. This can help us form expectations about how everyday lives can be influenced by future urban

[†]Singapore University of Technology and Design | lynette@sutd.edu.sg

[*]All characters are fictional, but their namesakes are partly inspired by my teammates. I thank Jude Kurniawan and Samuel Chng for their helpful comments when developing this chapter. Photo credit: https://unsplash.com.

43

SABREENA TRẦN

Age 12 (born 2088), Female
Student
High income
Ho Chi Minh City, Vietnam (pop. 15 mil)
Lives with her parents in a small, luxury apartment in a suburb

ABOUT

Sabreena is a genetically enhanced preteen, who is athletic, attractive, and intelligent. She enjoys a bespoke education, customized to her learning needs. She learns mostly from a digital archive through her avatar. Her parents send her to a social play group once a week in order to sustain some face-to-face human interaction. When she needs to travel, her primary means of getting around is by using her family's personal robo-land and aerial vehicle. Powered by sustainable energy sources, this sentient self-servicing vehicle, named "Sam", is able to plan its routes, pick up family members and run errands on demand (e.g. picking up parcels).

NEEDS

- Smooth traffic conditions in the city
- Her custom sneakers to be delivered to her location in under two hours

FRUSTRATIONS

- The family robocar "Sam" was hacked once and unavailable for two weeks
- Unable to go for her track training because of flooding in the city

Private — Public transport

Short — Long trips

SUMIT PATEL

Age 24 (born 2076), Male
Freelance Programmer
Low income
Bangalore, India (pop. 25 mil)
Lives in shared co-operative housing

ABOUT

Sumit is a university graduate majoring in bio-inspired computing. He takes part in the gig economy, completing small jobs designing algorithms. His work and leisure activities are mostly virtual. He plays a polyphonic pressure music instrument and performs live on virtual events with an international band. He eats and hangs out with his roommates at the neighborhood automated food district, which serves ready meals from ingredients that are restocked monthly. If he needs to travel, he relies on public transport, including the high-speed subway and on-demand robo-shuttle buses.

NEEDS

- Low monthly mobility service subscription fees
- Accurate predictions on public transport service disruptions

FRUSTRATIONS

- When the robo-taxi is disrupted by other unruly passengers or road users
- Heatwaves in Bangalore have been blamed for a recent mobility fee increase, due to added energy and transport system maintenance costs

Private — Public transport

Short — Long trips

LI CHAO

Age 60 (born 2040), Male
Part-time social worker
Middle income
Chongqing, China (pop. 9 mil)
Widower, lives alone in an off-grid
housing cell

ABOUT

Chao lives off-the-grid in a small housing cell which is part of a larger residential cluster. He moved to Chongqing from Shanghai 20 years ago after frequent flooding at the coastal city proved too disruptive. As a part-time healthcare social worker, he travels regularly to meet his patients. To get around the city, he often shares a ride in a robo-taxi. When in a hurry, he will rent a personal aerial vehicle with self-parking function. Other than meeting his patients, he does not need to travel much as he can pick up daily necessities and hot meals from a micro-fulfilment center located one block away from his home.

NEEDS

- Door-to-door, reliable public transport services
- To remain well connected to his digital life within the robo-vehicles

FRUSTRATIONS

- The micro-fulfilment center fails to anticipate the exact goods he demands
- Laments loss of privacy, since his movements and activities can be easily tracked

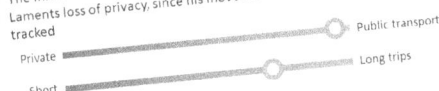

Private — Public transport

Short — Long trips

JUNE BHAI

Age 72 (born 2028), Female
Digital media editor
Middle income
Bandung, Indonesia (pop. 5 mil)
Single, lives alone in a senior-friendly
neighborhood

ABOUT

June migrated from Bangladesh to Indonesia when she was 24 years old. By delaying her retirement, she has continued to work as a digital media news editor. Her team comprises six journalists who are located all over the world. She works primarily from her home office, and occasionally walks to a co-working space to conduct enhanced holographic meetings. For her health and medical needs, she receives home-based care. She subscribes to a personalized plant-based nutrition plan, with minimal preparations required at her home kitchen. She enjoys volunteering in a nearby urban farm.

NEEDS

- Fast and reliable Internet communications channel
- On-time and on-demand deliveries to her home

FRUSTRATIONS

- Noisy road and air traffic near her home
- Urban delivery vehicles occupying curb space in her neighborhood

Private — Public transport

Short — Long trips

45

infrastructures. The storytelling allows empathy for city residents, allowing us to consider the social aspect of long-term transport infrastructure design and planning.

Biography: *Dr Lynette Cheah is an Associate Professor of Engineering Systems at the Singapore University of Technology and Design. She leads the Sustainable Urban Mobility research group, which develops data-driven models and tools to reduce the environmental impacts of passenger and urban freight transport. She is a Fellow with the Martin Family Society for Sustainability, and a Review Editor for the United Nations Intergovernmental Panel on Climate Change (IPCC) Sixth Assessment Report.*

— The Great Infrastructure Reckoning —

Mikhail V. Chester[*]

We had no choice but to reimagine how we built and managed our world. It was partly because we couldn't change the way we had been doing things; it was partly because technology outpaced our capacity to change; it was entirely out of our control and required us to come to grips with the fact that we were no longer in charge.

The signs were clear, but we struggled to see the approaching cliff. For decades, maybe centuries, infrastructure and the institutions that managed them had been able to deliver services in a market where the services that were demanded from them – water, electrons, automobility – barely changed. The institutions were heavily bureaucratic, with knowledge compartmentalized, significant barriers to making interdisciplinary collaboration happen, and creativity squandered. As such, the ability to innovate for the increasingly wicked and complex challenges that had emerged was handcuffed. Any little drive to restructure these institutions had been met with thunderous roadblocks as engineers, bureaucrats, politicians and the public were reluctant if not downright opposed to seed the conditions for innovation. It had become remarkably clear that innovation in all aspects of infrastructure was needed. Climate change driven extreme events and gradual change was shaking the roots of infrastructure design, infrastructure in much of the developed world was in dire need of rehabilitation and reimagination yet financing was largely uncertain, and the explosive potential of emerging and disruptive technologies including connected devices and artificial intelligence was accelerating and being incorporated into all of our dumb infrastructure systems and services. How we demanded services and understood our foundational infrastructure systems was about to radically change.

The car sped towards the cliff on cruise control with all of us on board. It was never that infrastructure were going to collapse and leave us with no basic or other services. It was a question of how if at all the institutions that were steering infrastructure would recognize that they would need to transition, the

[*]Arizona State University | mchester@asu.edu

47

preconditions necessary for the transition, and whether the transitioning would be proactive or reactive.

Costly Path

Today, at the dawn of the twenty-second century, it's clear how things got out of control. The institutions that managed infrastructure, the universities that educated future engineers and managers, politicians that steered financing, and the general public did not heed the warning signs. We continued to view infrastructure as a given, an amalgamation of hardware that needed the bare minimum to keep functioning.

Innovation Path

Today, at the dawn of the twenty-second century, the transitioning of how we manage and view infrastructure represents one of the greatest success stories of the last century. Early in the twenty-first century infrastructure agencies recognized that conditions were about to get far worse. For years they had kept their systems reliable and able to meet increasing demands but they had recognized the warning signs. Everyday decision about pump sizes, material design,

We handcuffed infrastructure managers in their ability to innovate. Politicians ignored the brutal reality that infrastructure has always and will always be designed around the environment. Infrastructure managers, with no training on cyber technologies and security continued to deploy software and hardware without a clear understanding of what it would mean for asymmetric warfare and artificial intelligence. Silicon Valley continued to generate and mine data on infrastructure services, fundamentally altering how the systems were used, without coordination with those who were planning or operating them. The results should have been obvious.

By the early twenty-first century infrastructure were already at the end of their design lives and these conditions led to a loss of reliability, escalating costs, and ultimately impacts across society. Climate change both through extreme events and gradual changes in background environmental conditions resulted in increasing failures and loss of services. Financing continued to be held hostage by increasing ideological polarization that tied funding to societal issues that we simply couldn't negotiate on. The large-scale deployment of sensors, microchips, software, and databases without any significant coordination or training on cyber security issues resulted in a plethora of opportunities for nation-states and hacktivists to stormwater management, and so on were becoming challenged by changing environmental conditions. They saw that the rate of technological change and connectivity in their systems was accelerating and how to manage these technologies was exceeding their expertise. They recognized that future generations would likely come to expect infrastructure to do very different things, many unpredictable at the time, and as such the underlying systems and organizations that manage them would need to be agile and flexible.

As educators of future leaders, universities recognized the impending crises and restructured their curricula to address the challenge. Basic proficiencies were expanded to include new competencies needed to address the challenge including cyber security and climate adaptation. Education was restructured around major societal challenges, and not the associated hardware.

At the same time infrastructure organizations, both public and private, recognized the need to restructure their institutions around these emerging problems and away from disciplinary silos. They recognized that the emerging challenges were interdisciplinary and required diverse teams of experts and the public. They became pioneers in harnessing emerging artificial intelligence programs to find solutions that had previously

hold infrastructure services hostage, thereby defining new forms of conflict that we had never seen. And the rapidly emerging technologies and their associated industries basically took over how infrastructure were used by taking advantage of a massive decentralized technology network that allowed their algorithms, whether intentional or not, to drive people towards use of infrastructure services in ways the infrastructure was never planned for, thereby creating myriad reliability challenges and unintended consequences.

Water and electrons kept flowing, mobility was still available, and information and communication technologies dominated, but in many ways we lost control of the underlying systems. The institutions that managed infrastructure simply weren't ready for these disruptions, independently or concurrently. They tried to keep up, at great expense, but their capabilities were limited as we didn't seed them with the resources to lead us through the transition into an increasingly complex world.

been inaccessible. The result was a radical transformation of infrastructure institutions, away from the bureaucratic divisional structure that had emerged within the industrial revolution towards new structures that merged human and artificial intelligence in ways that were much more accessible.

Agility and flexibility became central to these emerging systems. What had previously been rigid and locked-in hardware was reimagined as cyberphysical systems that were able to change quickly depending on changes in environmental conditions including climate, financing, demands, and technologies. Design paradigms that included safe-to-fail, deep uncertainty, hybrid gray-to-green, and hybrid centralized-decentralized became commonplace. The results were systems that were able to manage the consequences of failure much better, reorganize quicker and more effectively with changing environmental conditions, and integrate next generation technologies to the public's benefit at faster rates. We seeded infrastructure systems with the innovative capacities to adapt and transform.

The next century (2100 to 2199), like the last, offers remarkable new opportunities. And like the last century, positioning the infrastructure backbones of our society to give use the basic services we need to thrive will be paramount to our continued success.

Biography: *Mikhail Chester is an Associate Professor in Civil, Environmental, and Sustainable Engineering at Arizona State University where he runs a research laboratory focused on studying infrastructure resilience to climate change and uncertainty. His work spans a number of infrastructure systems (including power, water, and transportation) and hazards (including heat, precipitation, and wildfires). Chester is the co-leader of the Urban Resilience to Extremes Sustainability Research Network, a consortium of 10 cities across North and South America including roughly 200 academic and 150 practitioner partners in 19 institutions working together to develop adaptation strategies for urban infrastructure for extreme events.*

— Only connect —

Caitlin D. Cottrill[†]

> *"Cities have always offered anonymity, variety, and conjunction, quali-*
> *ties best basked in by walking: one does not have to go into the bakery or*
> *the fortune-teller's, only to know that one might. A city always contains*
> *more than any inhabitant can know, and a great city always makes the*
> *unknown and the possible spurs to the imagination."*

— Rebecca Solnit, Wanderlust: A History of Walking

We're meant to be somewhere, but there hasn't been a check in on any of the apps we share for a while. I'd say the phone was forgotten, but who forgets their phone? I had argued for the palm implant, but clumsiness was countered – the tech still isn't perfect, and shorts do happen. My usual eye-roll-and-huff didn't convince otherwise, so the old-fashioned unattached device it is. At least it's one of the good ones, and I've made sure that it's set to the City, rather than the old Town – it shouldn't matter, really, but the IP is still in contention, and not everyone has bought in to the wisdom of data sharing.

Idiots.

I check the map online - even without the phone (why did I agree to that?), the location implant should ping along the path, linking momentarily to the smart streetlights, the wireless, the cameras, the train, pollution monitors, traffic counters, kettles, pedometers, sewers, and all the other devices and systems that connect the City – but there's no trail of light showing the direction of travel. No social media, either. I searched the hell out of them before we hung out alone the first time, so I know all the handles, even the clumsily-hidden ones. IP addresses follow you. So does machine learning.

I said I wouldn't pry. Promised I wouldn't log in to the deeper tracks of bank transactions, private locations, face recognition, gait sensors, cryptocurrency

[†]University of Aberdeen | c.cottrill@abdn.ac.uk

trails, and behaviour-predicting algorithmic games. By providing me with the ability to do so, I was also gifted with the trust that I wouldn't.

But where... Not at home - I'm at home, now. And the phone is on the desk.

The City as a place of walking, wandering, and being lost is something I wish I had known. I don't want to be located today. Nothing bad, of course, I just don't want to have to go to the party or respond to the texts or answer for my whereabouts. I want to be alone, and unknown.

I grab a burner phone, just in case. I don't want my friends, or anyone, to be able to look at my regular apps and know where I am, but I want to know that I can contact them if need be. I bought this one with cash, braving the dodgy location and strange looks. I actually bought four of them at the same time. I don't usually carry paper money with me, especially as it's hard to get these days, so I made use of it while I had it. What else am I going to use it for? The ATM asked three times before it would hand over the bills – should I load this to your card? No. Shall I ping this directly to your implant? No. Direct pay an app? NO. Paper money, really? Yes, dammit, YES. I hope no-one saw me stabbing at the buttons so hard, or tracked the irritation in my staccato responses. The machine knows it, obviously, but I tried to pick one that says it doesn't share.

I have to plan to be unseen these days. I know the cameras and connected devices can see me as I walk through the City's streets, but I've done my best to map out paths that cross over territories. I hope that the IP agreements haven't yet been reached, and that my movements will blur at the barriers of data ownership. I turned off my personal chip location sensing, as well. I don't know if this does anything more than the futile pressing of the 'walk' signal, but hopefully it at least creates a legal barrier to tracking me. Hopefully, should anything happen, they'll be trapped in years of legal disputes to re-create my movements on this day. I hope they don't look at my browser history, where I created the map.

They will, though, if they need to.

Who are they? Not sure. The people who become involved when they need to, I suppose.

I'm not from here. I moved here, in part, to escape the visibility of the small Town. Where everybody knows your name, and your family, and your faults and foibles. The City seemingly offered an anonymous start – I could blend in to the background, shading my way against the backdrop of buildings and

54

shadows. But you carry your past with you, in the location trackers and beacons and social media histories. I tried to erase my digital traces, but the currency of bank accounts and rental histories, employment records and facial recognition holds true. The more we, and our data, connect, the less we're allowed to disconnect. The harder it becomes to grow and change and discard our previous incarnations.

I am not who I was, but my digital self, held in the records connected to the infrastructure across this connected City, won't let me forget. The City knows who I am and who I was, and will use that to assume who I will be. Just for today, I want to do my best to escape. To sit on the lakeshore as a random data point with no associated attributes, watching the gulls.

Biography: Dr. Caitlin D. Cottrill is a Senior Lecturer in the School of Engineering and Director of the Centre for Transport Research at the University of Aberdeen (Scotland). Her research focuses on applications of technology in public spaces, with particular focus on its' ability to contribute to more accessible, inclusive and reliable public and active transport networks. Her work in this area is complemented by research in data privacy and anonymity, particularly related to the impacts of always-on connectivity.

— *It is my nature* - Cape Town 2100 —

I wake to the sounds of shouts and wheelbarrows – the morning hustle to the market. I can't tell if it's the noise or the Cowrie™in my wrist – but I suppose that's the point of my "seamless transition into wakefulness." I kiss Efya on the cheek – she's scheduled for 6:30am – before telling the apartment to prepare a blast shower. Feeling refreshed, I order the Swahili Continental (ugali, beans, egg) for breakfast and meet the drone on the balcony – *7 minutes, they're getting faster!* I've been in the habit of large breakfasts – if you don't eat enough in the morning, you'll have a Cowrie™induced migraine by noon – it needs carbs too apparently. *For we are bound.*

I watch the traders below push in fresh mounds of heirloom tomatoes, onion, neon spinach, bananas – full containers, ready for market – it's tough going for them uphill, but whoops and whistles follow them as they surf the empty trolleys down the slope – how they stop at the port, I don't know – probably somewhat violently.

After breakfast, I slip on my bamboo jumper and head out. Early touts and business suits have joined the traders and I take my electric board up the hill to the market. The open streets are awash with stalls and vendors, smells and laughter. The occasional car sits rusting, claimed as a home or a staging zone for games or exposition. I pass a group of pastors in deep discussion, likely in strategy for the day's offering and goodly guidance.

The market may feel messy and chaotic, but it's the only place for fresh produce. Many avoid the bustle and the haggling, sticking to the freeze-dried meals shipped on glider-rail from upcountry. The market is full of the spoils brought in by the morning's wheelbarrows, but the streets are also lined with green walls of creepers and hanging pots offering sprigs of herbs, blue and purple tomatoes, curly beans, and granadillas. Their owners watch their space from the balconies above, calling their wares as customers take whole pots or harvest a few choice foods and wave their wrists to the owners in payment. When I reach my favourite zone in the market, a mini-orchard of about 12 banana palms,

[†]ICLEI Africa | paulkcurrie@gmail.com

Fanti sees me coming and pulls four just-ripe fruit from his current bunch and tosses them to me. I wave him our agreed price and catch his next missile of groundnuts wrapped in a transparent biosleeve.

With a midday snack to hold me over for the day, I walk to the Strand Station to catch the easterly glide-rail. The wind has calmed so I hope it will be operational. It's been offline for three days, keeping me from meeting my clients – I'd been booked to guide some visiting Costa Ricans, but the wind hadn't allowed a crossing for the last three days. The beach-side station has just been redone with mosaic inlay making it oddly appealing, if you accept their premise – the lost ruins home to new corals and kelp, with happy looking turtles and fish. They're trying to revitalize the tourist centre, where visitors come to scuba dive over the old harbour and waterfront district, or take pods around the Rolihlahla statue, now ankle deep in the sea. Robben Island disappeared sixty years ago, but the statue still stands – our Southern mirror of Lady Liberty, also dipping her toes in the Hudson, so I hear. The statue was supposed to be a global beacon for justice, though everyone claims Mandela for their own agenda – Pandas, Climate War, and our so-called Third Coming. Most have no clue that he was an anti-apartheid icon promoting an end to segregation through discourse and empathy. That history was washed away with the floods, and with the privatization of the G-NET. The floods did not wash away the inequality.

The glide-rail always fills me with wonder. First because of its sheer speed – nothing moves fast in the Mother city, especially since the end of automobiles – but mainly because of the view: the easterly glide-rail connects Devil Slopes to the mountainside city of Sir Lowry's Village. It was erected in the 70's a few years after the last lock broke and we lost the Flats. It represents our last tenuous grip to the mainland. On a clear day you can see the ruins below the water – today it's an azure green-blue affair, though I've seen it navy, royal, cerulean, dark, titan, cobalt and grey. The occasional roof peaks out, covered in algae, anemones and mussels, - convenient stop overs for wayward seals. People walked, worked and lived here. People dreamed here.

I do my dreaming in the mountains. I have a good life here, though occasionally I wonder about exploring beyond the island. The world feels far away and disconnected – our stories always seem to be twenty years out of date. But perhaps the G-NET can't actually keep up with everyone. The ruins become more frequent and visible as we near our destination. Here, the submerged city has a different look – people didn't seem convinced that the waters would rise so much - or so suddenly - and the homes show signs of waterproofing and blockades before they were abandoned – but these were the first innovations leading to the many floating villages that house the brave and the desperate. I can't imagine how they weather our winds and the waves.

The glide-rail slows and our carriage's pastor winds up his sermon with an incoherent attempt at poetry "… still thine fears and embrace your fellows, for nothing changes but how we navigate the winds." Still he gets a few murmured responses of "for we are bound."

I share my contact to a prospective client from Thailand, with a tilt of my hand, and find the exit of the station, sidestepping the crowd of touts and beggars hoping for something from the Cape visitors. I meet my clients four levels up at their hotel, now a rather kitsch spot hewn into the mountainside – but it's central and comfy. We wind down the levels, taking steps and the occasional funicular, to meet my pilot and his skip at the harbour. We pass the fishers casting their lines for the local restaurants and I wonder about their ratio of fish to historic flotsam. Our coastal tour explores the battlements of 2083, where the Cape tried again to secede from the Azanian Federation, claiming small-island statehood in a gambit to win climate reparations for its citizens. They were half successful, with a begrudging Federation subsidising the city's power, food, water and comms. The Federation must easily be making this up in tourist tax these days, given the swath of visitors attracted to see both history's greatest failure and greatest success in resilient planning. We are still vulnerable on our cliffside perch, and I think the Federation would prefer that residents moved inland, as has eventually happened to be a common occurrence for most coastal cities around the world – abandoned to the sea.

It was a satisfactory tour, with some bonus sightings of humpbacks among the seals and seagulls. A good tip means I can treat Efya tonight. I hop the glide-rail back to Cape Island and walk to meet her. I pause at a corner, taking in a street-tag I hadn't yet noticed – lasered on the wall is an image of a woman conjuring power and launching it – seemingly with her voice or being – towards me. I feel oddly emboldened, though I'm not sure for what. Slanted next to her is a mantra:

I've seen it all
time sees all
those blips
and lengths
of love
and terror
as terraformed
changed
I do not fear
change
it is my nature

This floats in my mind as I continue walking, with a familiar sense that nothing changes – nothing of substance anyway – I can never really articulate if this is a depressive or emancipating feeling – it just is. *For we are bound.*

I meet Efya waiting at her office entrance and we stroll up signal hill to watch the sun set and the city lights come on. This is one of the last open places with green space in the city. Something about the collective unconscious of the city needed it to remain a gathering area – while hanging homes stretched farther up the cliffs of Table Mountain and Devil's Peak, supported by a tangle or mini-glide-rails, funiculars and elevators. The Lion has remained untouched.

The green hue of city light meets the starry sky – Jupiter and Mars have been easily visible for the last month. I wonder how they feel about the exploits of our city – how do they feel its shifts and yearnings? *Salduro* from nearby speakers breaks my reverie and I lead Efya towards the beats. Around us, other friends and couples enjoy the dance, with sundowners in their hands, or speaking absently into the G-NET.

Tonight, I know I'll dream of the azure water – it surrounds and engulfs me. But I'm sure my Cowrie™will block the nightmares. *For we are bound.*

I'VE SEEN IT ALL
TIME SEES ALL
THOSE BLIPS
AND LENGTHS
OF LOVE
AND TERROR
AS TERRAFORMED
CHANGED
I DO NOT FEAR
CHANGE
IT IS MY NATURE

Biography: Paul is fascinated by the interdependent relationships which shape African urban systems and give each city their unique flavour. He uses urban metabolism, resource nexus, political ecology, storytelling and photography as conceptual and literal lenses for shaping sustainable, inclusive cities. At ICLEI Africa, he supports local governments to apply systems perspectives to food, water, energy and nature, and facilitates the development of policies and plans appropriate to contexts of rapid change and uncertainty.

— The City of 2100: An Idealized Look —

Cliff I. Davidson

Introduction

What will cities look like in the year 2100? Many authors have explored this question and, as expected, there are wide ranging opinions. Based on history, what we know of science and technology, and the current world population, there are some changes we might speculate on. The ideas which follow are based on an optimistic scenario: We assume, given our current knowledge, that future government leaders will make decisions that benefit all of society, resulting in a high quality of life.

Climate change is expected to accelerate by the end of the century, bringing warmer temperatures and storms with increased intensity. Clean water is going to be highly valued and used for essential needs; we are not likely to be using potable water in our toilets. Our energy portfolio will have shifted away from fossil fuels: solar, wind, and other renewables will likely dominate. The world will be predominantly urban.

Much like today, myriad factors will cause cities to change; people will move to where the jobs are and they will choose residences near or far from work, based on what they can afford as well as social and environmental preferences. These decisions will, in turn, affect climate through carbon emissions resulting from lifestyle choices. But the reverse is also true: changes in climate will influence where people live and how they spend their time. Thus, urban growth will influence and be influenced by climate change.

Models have been created to predict how cities change over time. Many of these predictions, designed for use by urban planners, are restricted to one or two decades. This is the longest time period over which regulations and city operations are typically expected to remain constant. But recent efforts have involved systemic modelling, the combining of models and knowledge bases from various disciplines, to identify possible long-term changes. For example, to study urban change over the coming century, Masson et al. (2014) propose

†Syracuse University | davidson@syr.edu

linking models from multiple domains: use of a socioeconomic model to account for changes in economic growth or decline in a city, use of a geographic model to identify locations of urban development, use of a building energy model to account for HVAC operation by residents responding to a hotter climate, and use of other models to account for the many factors that affect the complex nature of urban change.

A rigorous analysis of these models and others to describe possible urban infrastructure changes by 2100 is beyond the scope of this chapter. Rather, we will adopt some ideas from the literature to surmise what cities some eighty years from now may look like, given what we know about attempts to live sustainably today.

Urban Design in 2100

The imagined "City of 2100" is located in the Northeast region of the United States where there are four seasons, about one meter of precipitation on average distributed roughly evenly throughout the year (a fraction of this is snow), enough land for farming, and close enough to the Great Lakes to fish during much of the year. We follow some of the ideas of Revi et al. (2006), who reported on the sustainable redesign of the city of Panjim in the state of Goa, India. Revi and his coworkers chose this city of around 115,000 (Census of India, 2011) to conduct food, water, energy, and transportation analysis. They reported that the existing land area of the Panjim metropolitan region, 170 km2, would be sufficient for the population to live in a sustainable way without inflicting ecological damage. This would require strategic land use changes, especially for growing fruit trees and rice in this tropical climate. Since Panjim is on the coast, fishing could provide additional food. Based on the modeling of Revi's team, existing natural ecosystems throughout the 170 km2 area could continue to thrive with 115,000 people living off the land and the nearby ocean.

The concept of an "autarchy", a sustainable city that is self-sufficient within its borders, dates back to the ancient Greek city states in the 4th and 5th centuries BCE as discussed in the writings of Plato and Aristotle (Hughes et al., 2001, p. 61-62). Interestingly, one government controlled the built-up urban area of Athens and its surrounding farmland and forests out to the boundaries of the city state. The population of ancient Athens was around 100,000 and not much larger, similar to the maximum population of other city states in ancient Greece. Although the ancients certainly did not live sustainably (i.e., autarchy was never achieved: deforestation was a major problem, and some food was imported from distant colonies), the availability of resources in the surrounding countryside was probably one factor limiting their population, in addition to the battles with other city states. In this paper, we assume a population in the range 100,000-200,000 for our proposed sustainable City of 2100.

Current practices in planning consider urban spaces first, and allow developers to propose residential, commercial, and industrial areas designed to maximize economic output. Parks and other green spaces are then implemented where no urban development has occurred. In contrast, we propose to consider the natural landscape prior to urban development first, examining the ecological productivity and the flows of water and energy through the natural ecosystem. We can then consider where urban development would minimize damage to the ecosystem while providing an acceptable quality of life for people.

Figure 1. Layout of the City of 2100 in idealized form. Individual built up areas of the city, or settlements, are shown as small gray circles. The rest of the area is farms and protected ecosystems.

We propose a city composed of perhaps ten or twelve settlements in rings around government buildings in the center, as shown in Figure 1. Each settlement would have its own neighborhood council that could propose decisions for approval by the central city authority. The space between the settlements would contain protected natural ecosystems as well as farms, such that the city land could provide a sizeable fraction of the food and water needed for inhabitants. Wind farms and solar panels are proposed on appropriate sections of the land. Instead of isolated green spaces within a sprawling city, there will be several small but dense urban settlements embedded within a larger area for farms and ecosystems, a concept suggested by Revi et al. (2006). Figure 1 is an idealization of the proposed city. It would be reasonable to establish property

65

lines between settlements consistent with ecological boundaries, such as watersheds, so local decisions can prioritize the preservation of ecosystems. Using ecological boundaries was suggested by Lewis Mumford (1970, p. 408), who believed that such governance could help people with the challenging task of living within limits set by nature.

The population of the city would depend on the productivity of the farmland within the city boundaries, the amount of fish that could be provided sustainably from the Great Lakes, the preferred diet of city residents, and natural geographic constraints. For comparison, Revi et al. (2006) calculated the desired land use fractions for Panjim in 2100 as 25% forest, 21% food production (rice and fruit) with fish available from the ocean, and 6% built up urban land for the settlements. Other proposed major fractions of land include mangroves, grassland, and the existing river flowing through the city. Assuming invariant population over time, this implies a density of 110 people/hectare within a settlement. Revi et al. proposed target population densities of 150-300 people/hectare. The settlement areas in Figure 1 are also about 6% of the total area of the hypothetical City of 2100.

Revi and his team considered the energy that could be produced by renewable sources and concluded the Panjim metro area could provide 1500 watts per person, which would be enough to allow comfortable lifestyles. It is worth noting that this is far less than the U.S. per capita energy consumption by a factor of eight. Providing sufficient water in Panjim would be a challenge, in part because the annual rainfall of about 3 m occurs during the monsoon season, with much less rainfall at other times of the year. In Northeast U.S., where the rainfall is about one meter per year, the challenge would be even greater.

Hoornweg (2015) notes that cities of the future must have sufficient autonomy to define their own pathways to sustainability as well as their proposed sustainability endpoints. This greatly increases the challenges of governance, especially during the transition years. As the City of 2100 plans for the future, its leaders in government, business, and other institutions will need to connect with leaders of other urban areas and negotiate for materials, information, and even necessities of life to ensure the population's needs are met. The constraints of sustainability also add complexity to urban development decisions in steady state, such as attracting companies to relocate. Achieving a balance between federal authority and local authority will be another challenge as the autonomy of municipal governments grows. Connectivity with other cities will be crucial since negotiations will be ongoing as climate, education level, economic conditions, and demographics change over time. In the idealized world of 2100, there are no long-term changes in the population of most countries, although the population of individual cities will fluctuate as the elderly die and young people seek employment.

Figure 2. Center of a settlement in the City of 2100. High speed trains, roads, and people movers as well as sidewalks are shown.

Figure 2 shows the center of a settlement. Sufficient land exists so that residential buildings as well as commercial and office buildings need be no higher than 5-6 stories to achieve population densities in the range above. Thus there are no high-rise buildings. All space is used effectively. Each settlement is compact, permitting bicycling or walking, but residents may also use people-movers (small railcars shown in the figure) to travel quickly across the settlement. No personal cars are needed; buses, taxis, vans, and trucks use roads to travel within one settlement or to neighboring settlements in the city. The high-speed train powered by renewable energy transports people and goods quickly from one settlement to another within the city and to other cities. All modes of transit shown, except the people-mover, continue from one settlement to the next, so people can easily reach neighboring settlements.

The design of infrastructure, shown in Figure 2, follows the Envision Rating System of the Institute for Sustainable Infrastructure and the American Society of Civil Engineers (ISI, 2020). Envision provides a maximum of 64 credits in rating sustainable infrastructure across five categories: Quality of Life, Leadership, Resource Allocation, Natural World, and Climate and Risk. All three pillars of sustainability, Environmental, Social, and Economic, are incorporated into this rating system. For example, the infrastructure is designed to specifically improve mobility and access for the community, while promoting public health and safety. Residents have been involved in the process of planning the city, and therefore the city infrastructure integrates well with the activities of the community. Use of materials and energy has been minimized in both construction and operation of the infrastructure. Bodies of water and natural plants and animals are not harmed in the building and operation of the urban infrastructure, and emissions of greenhouse gases have been minimized.

Figure 3. View of a building roof in a settlement, showing wind and solar energy generation as well as roof gardens for growing vegetables and transient storage of stormwater.

Figure 3 illustrates buildings in the City of 2100 and the capability they have to produce wind and solar energy. The wind generators can receive the full force of the wind since all buildings are of similar height, thus avoiding the problem of wind blockage caused by buildings of varying heights. Two types of wind generators, shown in Figure 3, include tall oscillating cylinders, i.e., vortex turbines, and gyroscopic wind generators. In addition, solar panels receive the maximum amount of sunlight without being shaded by nearby buildings. Many of the buildings have green roofs to capture stormwater, thereby minimizing the potential for flooding. The green roof shown in the foreground also includes some trees and vegetable gardens. Besides green roofs, the widespread use of other forms of green infrastructure can help manage stormwater, such as rain gardens, street trees, bioswales, constructed wetlands, and permeable pavement.

Discussions and Conclusions

In this paper, we have imagined an idealized Sustainable City of 2100. We assume humanity has achieved an understanding of the importance of the natural world in supporting all forms of life on the planet. Rather than ignore the current ecological crisis, or dismiss its importance, we imagine that people have risen to the challenge and created a new system of sustainable cities that are able to support human life in perpetuity. Can this really happen?

Many cities around the world are collecting data on resource use, waste discharges, biodiversity, economic productivity, housing, public health, and numerous other metrics of sustainability. The Global Platform for Sustainable Cities (GPSC) provides assistance so cities can choose their appropriate metrics, rate themselves on their current position, and establish plans to transition toward sustainability (GPSC, 2020). More than two dozen cities are working with the GPSC, including some large cities such as Lima, Johannesburg, Jaipur, Beijing, and others.

There are also efforts to help civil and environmental engineers design infrastructure that is more sustainable. For example, the ASCE book Engineering for Sustainable Communities by Kelly et al. (2017) provides valuable information on transitioning the engineering community toward sustainability by considering topics such as transportation, water resources, energy, waste management, climate change, project management, working with communities, and a number of specific case studies in various cities. Thus, education of engineers in sustainable infrastructure is of great importance.

The City of 2100 described here is hypothetical, and represents one of many designs that could exist for a sustainable city. In today's world, numerous small "ecovillages" (pilot projects) have actually been built, ranging in size from a few households to hundreds of people or more. One of the longest surviving ecovillages is Gaviotas, Colombia, founded in 1971 (Weisman, 1998). The Global Ecovillage Network (2020) includes about 10,000 contemporary communities around the world, sharing practices that promote sustainability.

Despite many small-scale efforts around the world, our planet is in danger. Climate change, loss of biodiversity, increasing use of resources, and other dimensions of global change due to human activities are occurring faster than we can address them. Our priorities must become focused on confronting these changes in substantive ways, and at scale. To do so is an urgent need. We cannot afford to wait.

Biography: Cliff I. Davidson is the Thomas and Colleen Wilmot Professor of Engineering, and Director of the Environmental Engineering Undergraduate and Graduate Programs in the College of Engineering and Computer Science at Syracuse University. He is also the Director of the Center for Sustainable Engineering at the University, and is an affiliate faculty with the Syracuse Center of Excellence in Environmental and Energy Systems (CoE). He is a Faculty Fellow at the Syracuse CoE, and is a Fellow of the American Association for Aerosol Research, the Association of Environmental Engineering and Science Professors, and the American Society of Civil Engineers.

Acknowledgements: The author gratefully acknowledges the efforts of graduate student Monisha Arnold in the School of Architecture at Syracuse University for producing the drawings in this paper.

References:

Census of India, https://www.census2011.co.in/census/metropolitan/414-panaji.html, accessed January 1, 2020.

Global Ecovillage Network, https://ecovillage.org/, accessed January 1, 2020.

Global Platform for Sustainable Cities, https://www.thegpsc.org/usf, accessed January 1, 2020.

Hoornweg, Daniel, *A Cities Approach to Sustainability*, Thesis, Doctor of Philosophy, Department of Civil Engineering, University of Toronto, 2015. website link, accessed January 1, 2020.

Hughes, J. Donald, *An Environmental History of the World: Humankind's Changing Role in the Community of Life*, Routledge Publishing, London and New York, 2001.

Institute for Sustainable Infrastructure, https://sustainableinfrastructure.org/, accessed January 1, 2020.

Kelly, William E., Barbara Luke, and Richard N. Wright, *Engineering for Sustainable Communities*: *Principles and Practices*, ASCE Press, Reston, Virginia, 2017.

Masson, V., C. Marchadier, L. Adolphe, R. Aguejdad, P. Avner, M. Bonhomme, G. Bretagne, X. Briottet, B. Bueno, C. de Munck, O. Doukari, S. Hallegatte, J. Hidalgo, T. Houet, J. Le Bras, A. Lemonsu, N. Long, M.-P. Moine, T. Morel, L. Nolorgues, G. Pigeon, J.-L. Salagnac, V. Viguié, K. Zibouche. Adapting cities to climate change: A systemic modelling approach. *Urban Climate*, Vol. 10, pp. 407-429, 2014.

Mumford, Lewis, *The Myth of the Machine*: *II. The Pentagon of Power*, Harcourt Brace Jovanovich, New York, 1970.

Revi, Aromar, Sanjay Prakash, Rahul Mehrotra, G.K. Bhat, Kapil Gupta, and

Rahul Gore, Goa 2100: the transition to a sustainable "RUrban" design, *Environment and Urbanization*, Vol. 18, Number 1, pp. 51-65, 2006.

Weisman, Alan, Gaviotas, *A Village to Reinvent the World*, Chelsea Green Publishing, White River Junction, Vermont, 1998.

Biography: Cliff I. Davidson is the Thomas and Colleen Wilmot Professor of Engineering, and Director of the Environmental Engineering Undergraduate and Graduate Programs in the College of Engineering and Computer Science at Syracuse University. He is also the Director of the Center for Sustainable Engineering at the University, and is an affiliate faculty with the Syracuse Center of Excellence in Environmental and Energy Systems (CoE). He is a Faculty Fellow at the Syracuse CoE, and is a Fellow of the American Association for Aerosol Research, the Association of Environmental Engineering and Science Professors, and the American Society of Civil Engineers.

— Peace Day in Saint Pierre and Miquelon – July 14, 2100 —

Sybil Derrible[†]

Saint Pierre (Saint Pierre and Miquelon), July 14, 2100

Finally, it's Peace Day! My body went from floating in the cosmos to being fully aware in a fraction of a second. Automatically and effortlessly, as I open my eyes, my pupils contract, and as my ears sharpen, I hear faint but familiar voices in the kitchen. Having barely gained consciousness, my heart starts pounding, my muscles tense up, and my body mechanically starts rising and before I know it, I am seating up in my bed. I am so excited. Finally, it's Peace Day!

I get up, activate the command to untint the window.... wow, a great blue sky. The perfect weather to hang out outside with friends. I can feel a little breeze as the air exchanger fills the room with fresh oxygen to fill me up with energy, but it's useless today, I am already full of energy.

I stretch my arms and body and leave my bedroom, aiming for the kitchen. My father is standing in the middle of the room, looking a bit tired, with his shoulders slouched forward a little. He is arguing with my mother, who is by the counter holding a cup. Apparently, the UV water treatment system in the basement needs to be fixed, but my father says it is nothing like the water leak we had two years ago from the solar water heater in the attic. My mother reminds him that he needs to get ready for the gathering at 11am. The entire city council will be there and, as a counselor, she wants to make sure they arrive on time. As she says those words, her back straightens up, her face becomes more serious, and a small wrinkle appears on her forehead. It is especially important this year because she spearheaded the campaign to get a new energy storage facility for the microgrid of the neighborhood where we live, and she knows residents from other neighborhoods will ask her about it. Apparently this one is ultra-low power and more resilient because it can adjust to any power demand while

---

[†]University of Illinois at Chicago | derrible@uic.edu

73

being connected to the main grid and without affecting our home energy storage receivers, even if it shuts down. I am not sure what it means, but she looks stressed and she says she wants to make sure she gives a good impression during the gathering.

My father agrees, grunts a bit, picks his cup of coffee and finishes it, and heads downstairs, almost bumping into me on the way out of the kitchen.

When my mother sees me and sees how excited I am, her face relaxes. She gives me a kiss on the forehead, tells me that they already finished the baguette but she will order a fresh one from the baker and she asks me if I want anything special.

"Petit pain au chocolat!" I immediately respond without blinking or even thinking for that matter.

She smiles—she knows it's my favorite—she then taps something on her dev, and tells me the baker's delivery oto will get here in less than ten minutes. The drinks dispenser is on, so she tells me I can pour myself a hot chocolate. She then leaves to go take a shower.

In the living room, my older sister is watching the telereality (TR). Her body seems comfortably seated in the couch, but her neck is extended and thrusted forward, chin first, as if being a few centimeters closer to the TR provided a better experience. She barely blinks and is completely captivated by the show. I am not surprised, they are broadcasting live the Paris peace march, featuring many testimonies from living war survivors. Every year, the event is watched by everyone, but it is even more important this year since representatives from more than a hundred countries were invited and are attending the event to commemorate the end of the war. My sister learned about the war in her class last year and she has been obsessed about it ever since.

A few minutes later, as my hot chocolate is being poured in my mug, the fresh baguette and the petit pain au chocolat arrive... mmm, I love the smell of freshly baked goods. I sit at the kitchen table and watch the TR from afar. I am ecstatic because Marie is attending the Paris peace march as well. She is the best voyager of all time. I think she has been everywhere in the world, met an impressive number of different people, and encountered many different cultures. She has an amazing show and I have watched every season with my family. When I am older, I want to travel the world like her.

My older brother comes in, half asleep. He slumps in the second couch and changes the TR stream. My sister and I get mad at him... again. With a hoarse voice, my brother says, "I'm the oldest, I do what I want."

We start calling our parents and we hear dad yelling from the basement.

"Luke, leave them alone, will you, go watch your TR in your room. You just got a new one for your birthday."

Luke mumbles something unintelligible, gets up, and drags his feet out of the living room. After a few seconds, we can hear the show he wanted to watch blast out of his bedroom... so typical of him, always trying to annoy us. My sister switches the stream back to the peace march, she increases the outside noise canceling system to a specific frequency so as not to hear my brother's TR, and we both watch the march, being extra quiet when they interview Marie (my sister wants to travel the world like Marie too).

After a few minutes, I hear my mother shouting:

"Pascal! The water! What did you do?"

"Sorry honey, I had to turn off the water for just a minute to change the UV curator. Give me a sec." my father replies.

"Hurry up" mom says, "I have shampoo in my eyes. I told you not to do this today. We don't even need a UV curator anymore anyway."

"I know, but I sell them, so I need to test them, and I am trying a new one now" my father says, and after a few seconds, he shouts: "All finished, try again, honey."

We then hear the shower running again.

At 10am, the Paris peace march is over. It's then my turn to get into the shower and wash up. I love the new soap we have; the bottle says it's all natural but it's got some organic nanoparticles to be extra soapy and to smell good all day. Once done, I go back to my bedroom, close the door, dress up, and pick my favorite pair of carelies; I want to look great in front of my friends, especially today. Through the door, I then hear my mother telling my father to hurry as he shuts the bathroom door to take a shower.

At 10:30, my friend Max calls me over the dev and asks: "Are you ready to go to the square?"

I ask my parents if I can go and they agree. I tell Max:

"Yes, pick me up when you are ready."

"Ok, I will be there in two minutes," Max replies.

I go out and prepare my new MM-3 bicycle that is made of strong, ultra-light material that is also used to make all-terrain otos for voyagers. It is like riding a feather. Plus, the steering wheel is made of thermochromic light-sensitive material, and it therefore changes color based on the temperature and luminosity. With 24° Celsius and a bright blue sky, the wheel is mat yellow today. Tuyet!

Once I am barely ready, Max pulls up with a MM-3 as well and brakes to a skidded stop next to me, looking proud with a straight back and wearing sunglasses.

"Ready to go?" Max says.

"Let's go," I reply.

And we both pedal as fast as we can to get there as rapidly as possible. People are getting ready for the 11am gathering and, as a result, there is quite a bit of traffic even though today is a holiday, but not enough to slow us down. We swerve around all the otos. As we get closer to the square and as the traffic increases, one of them brakes brusquely, and we hear the driver yell:

"Be careful with your bikes now. Slow down. You're lucky this oto's braking system works so well or you could have gotten yourselves killed."

"Sorry," we both yell at the same time, but we keep going as fast as we can, until we get there.

The whole trip from home to the main square could not have taken more than five minutes (maybe a new personal record!). We see a group of friends at one end of the square and join them. Some of them are sitting down on the sidewalk, some are sitting on their bikes, and some are standing. They all have a fairly serious look; I bet they were comparing their bikes to try to determine who has the best.

"Nice rides," one of the them says.

"Thanks," we reply.

"Are those MM-3 bikes?" We hear.

"They sure are," we respond proudly.

"They're ok," another scoffed, "but they also don't look as nice as my brother's XG-5. It's going to be mine in less than six months, you know."

Max and I look at ourselves thinking the same thing... she is just jealous because she only has a MM-1.

The main square is decorated for Peace Day. Posters, flags, and buntings are placed everywhere, displaying the peace colors, all made of recycled bioplastics. There must be about thirty stands as well, some of them will host games, others will sell peace memorabilia, and many will also sell food and drinks. I am personally looking forward to having a big ham and butter sandwich for lunch.

As we are getting closer to 11am, people are starting to set up in the center of the square and are chatting. We are still a bit further away, but the crowd gets larger, to maybe five hundred people. As the noise level from the crowed increases, we actually have trouble hearing ourselves. At 10:55, I see my parents arrive, both panting a bit. My mother gives a quick frown to my father, whose shirt is a bit untucked out of his pants. She then seamlessly turns around, gives a big smile to everyone else, and delicately gets through the crowd to be closer to the front.

"Hi Carol, hi Pascal," I can hear some of them say. "Happy Peace Day."

"Happy Peace Day," my parents reply.

Soon after, the mayor starts her speech. She is relatively petite and has a soft and calm voice, but she carries a strong presence, and as soon as she starts talking, the crowd becomes silent. She has short hair, high heels, and she is wearing a suit as is proper for the occasion. In her speech, she first recalls the history and culture of the archipelago, how we were involved in the wars of the past two centuries, how we had to adapt to climate change, and how we managed to overcome these challenges by sticking together—it is remarkable how petty fights and differences between political leanings and religious views simply dissolve in the presence of a common and imminent threat as if they had never existed in the first place (no matter the threat, we are always stronger together than apart). A smile comes to her face when she talks about local dishes, listing several locally-grown produce including zucchini, tomato, leek, potato, and cabbage, as well as some wild berries, including cloudberries (my personal favorite), along with local fish and seafood, not forgetting to mention the delicious deer and rabbit meat hunted locally—for example to make paté—and explains how food enabled us to reconnect with the traditional cuisine that her great-great-great-great-great-great grandmother used to cook (I might have omitted or added an extra 'great;' it's hard to keep count after repeating the same word a few times). She then talks about how the population increased after the last war, and how delighted we were to welcome so many new people, looking fondly at her husband whose family arrived at the time, and how Saint Pierre and Miquelon has always been a welcoming place that has only become stronger as it has become more diverse. This is something to which most of us can relate. I personally feel proud as my grandfather served in the war and came to Saint Pierre to train the local population, and eventually decided to stay here.

Mayor Park then talked about how much the city had changed in the twentieth century, how the houses are still painted with vibrant colors, but they are incredibly well insulated, how the city has had to adapt cleverly to sea-level rise, and how we manage to generate our own electricity and run our own system of reliable microgrids. She also proudly tells everyone how we do not need home-scale water treatment systems anymore since the water that is distributed is naturally cleaned by the bio-reservoir, and water pipes are coated with self-treating biofilms, although some people still use a UV system, using air quotes as she says "to be safe" (at this moment, I look at my father and see his cheeks flushing a little, but I am not sure if it is because he is embarrassed or angry). She then talks about how we barely generate any solid waste anymore and recover everything from it anyway, and how we have become an example for Europe since we had become not only energy and carbon neutral, but we have actually become carbon positive thanks to our healthy boreal forest. Pointing to a small group of people not far from her, she recalls everyone how delighted we are to host a delegation from the Isle of Bute in the Republic of Scotland. They are here to learn from us, and to thank us for our time, they have brought a few cases of Scotch for all of us to enjoy on Peace Day. She then talks about all the great things that the twenty-second century will bring and she invites everyone to stick around today and enjoy tonight's fireworks that should be spectacular if the weather permits.

At this point, the mayor stops her speech as a few municipal employees dressed casually distribute clear paper cups with Scotch or red wine for those who drink alcohol. Because I am below the legal drinking age, I am given a cup of cloudberry juice, which I believe must be much better than Scotch or wine (although I haven't really tried alcohol yet). Once everyone has a cup, Mayor Park picks up the microphone and cries "Happy Peace Day" and everyone loudly shouts "Happy Peace Day" before taking a sip. Despite the rising chatter from the crowd, I am still savoring my cloudberry juice, feeling a few goose bumps and shivers down my spine that are spreading symmetrically through my body all the way to my toes as I am swallowing the nectar.

Once done, municipal employees go around again to pick up the cups for reuse or recycling. Personally, I look around and walk toward my parents. My mother barely touched her drink, while my father's cup is nearly empty. As I get closer, I see that they are talking to an older couple and I hear one of them say:

"Carol, as a counselor, you really ought to do something about kids on their bikes. As we drove here, two of them zoomed past us and the oto had to stop suddenly, almost giving me a heart attack...."

My heart rate shot up! Without thinking, I turn around, retracting my head in my shoulders like a turtle, and I meander through the crowd so they cannot see me. On my way, I pick up Max and we walk towards another group of friends. While I am still upset, I hear a voice say:

"I like you carelies."

"Thanks," I reply. My heart rate slows down, and within what must be less than second, I come back to my normal self, completely forget what has just happened, and respond:

"I got them on a trip to Chicago last April over the break." Alexis, who commented on my carelies in the first place, then added:

"Chicago. Tuyet!"

"Yes," I reply. "It's the largest city in the U.S. I learned that it transformed itself in the mid twenty-first century when people started flocking there because of sea-level rise on the coasts. What a great place! I loved it and I want to live there when I am older."

The conversation then continues on history and geography for a few minutes, mostly about the topics that we learned in class last year. There are six of us at first, standing by the fountain, but two more friends joined us soon after. We each talk about our summer plans. Many are actually back from Langlade for a few days and will go back soon. I am bit jealous. Langlade is the island next to Saint Pierre and linked to Miquelon by an isthmus. A lot of people from Saint Pierre have summer homes in Langlade. Some people have settled there permanently as well. It is a twenty minute water-oto ride from Saint Pierre or a thirty minute land-oto ride from Miquelon, that is itself about a one hour water-oto ride or a ten minute sky-oto from Saint Pierre.

Because Langlade has no power grid or water distribution system, all the houses are completely self-sufficient. Some of them have solar panels but many simply have small energy storage units that they charge once in a while since they do not consume much energy anyway. For water, they tend to have wells in their backyards that they line with biofilms and they also put a living biochar in the bottom to treat the water, and in the end, it is completely safe to drink. All the food waste is used to make compost for private and public gardens or put in the common digester. All the paper packaging is burned, for example for fun in the evening (campfire), or used to heat houses and cook food in old-style wood burning stoves, or it is disposed of in the digester as well. Only the electronics are disposed of in Miquelon or in Saint Pierre to be recycled, everything else is recovered on site like in most places. I have learned all of these things from my dad.

As we are getting closer to lunch time, the air is getting filled with wonderful smells of pork and lamb sausages and chicken being grilled. While some of the people are leaving, likely to eat at home, the remaining crowd is getting closer to the food and drink stands. My stomach starts rumbling. I go to my parents who are still talking with a few people (I first check to ensure the older couple from before left; they did). When I arrive, I hear my mother talking about the new neighborhood energy storage facility. Her face bears a clear smile of pride, although it is a bit tensed as well. I can tell that she is paying attention to the reaction of the people to whom she speaks to measure their level of interest. I therefore go to my father, who has been quiet, and ask him for money to buy a sandwich:

"Are you sure you don't want to have lunch at home with us? We are leaving soon," he says, with a tone and an expression that suggests that he has been bored for a while.

"Can I eat here and stay with my friends?" I ask back.

"Sure, that is fine, but ping us or come home if you need anything, ok?" and he transfers some money on my dev and says, "That should be enough to buy yourself at least two ham-butter sandwiches. I know you have been thinking about it for days."

I laugh and then tell my father, "Dad, Elliott also told me that the water from their well in Langlade tastes even better than the water here in Saint Pierre and they use biofilm, living biochar, and a UV curator from our store."

My father pats my head, smiles, and responds, "Of course, we only sell great and reliable products. I am not surprised. Plus, it's always good to be off the grid if you can."

"Exactly," I reply. "We even learned it at school."

I give him a big smile, turn around, and run to my friends. We then all walk together to the food stands, where I order a ham-butter sandwhich that ends up being as delicious as I anticipated.

The entire afternoon is then spent having fun, playing games, running around, and riding our bikes. In fact, I won a two kilometer bike race with my MM-3; as I said, it's like riding a feather.

By far, my favorite game is the world peace game. The rules of the game seem complicated at first, but they are not once you have played a few times. Essentially, all of us are separated into several groups; today, we are 20 people divided into five groups of four. Each group starts with zero points, and the goal is to gain points by meeting some objectives, without decreasing the points of other groups (if done well, meeting an objective can in fact increase the points of other groups). All groups have different objectives to meet. The final goal is to come up with strategies to maximize the number of points of one's own group while maximizing the points of all groups as well. The game is highly strategic and requires collaboration between all groups. In the end, the groups that win need to meet two conditions (we call them Wardrop's conditions):

1. The number of points of the group would not be higher with another strategy, and

2. The sum of the points of all groups would not be higher with another strategy.

The goal is therefore not to have more points than other groups, but to do as well as possible without negatively affecting the other groups. The game is difficult, and some people simply do not want to collaborate. Today, the world peace game was about operating infrastructure. The five groups were divided in five infrastructure service sectors: electricity, water, transport, telecommunications, and solid waste. Each group was responsible for one infrastructure system individually. The objective was to properly provide a service, but because all sectors depend on one another, collaboration between the groups was needed. For example, if the water group did not collaborate with the transport group, then they could not repair their water conduits and they would therefore not meet their objective. I was part of the water group, and as the child of an engineer, I thrived during the game. Out of the five groups, three groups won (including mine of course). The two groups that lost came up with strategies that would give them higher points, but that would jeopardize all other groups, thus making all groups worse overall.

By 4:30pm, most of us are exhausted and I decide to head home. Max already left 30 minutes ago, in a fury. Max's group is one of the two that lost and the team members disagreed extensively during the game; no team, company, or even society for that matter, can thrive without clear communication and compromise. As the old African proverb goes: "If you want to go fast, go alone. If you want to go far, go together." I therefore bid farewell to everyone and tell them that I will meet them a bit before 9pm for the fireworks. I hop on my bike and pedal back home, taking much more than the five minutes it took me to get here this morning since I am exhausted, even though there is no traffic whatsoever at this time.

When I get home, the house is empty. Both my brother and my sister are at the main square (I saw them earlier). My parents are gone as well. I ping them on my dev and learn that they are having tea with the Scottish delegation. I hate being alone at home. No matter how big or sophisticated a house is, an empty house is a house without a soul. A technology will never replace the energy and warmth that a human being can bring to a space. In fact, no matter how much one loves a technology, the technology will never love them back. In the end, technology is only a support to help bring people together and exchange—a simple accessory. With my eyes half-closed already, I head to my bedroom and lie on my bed to relax. I want to take a small ten-minute nap.

One and a half hour later, I wake up in a panic. What time is it? Did I miss the fireworks? I can hear some noise, unconsciously bringing me to smile since it means I am not alone in the house anymore. I walk to the living room and see that my parents are back. They are both watching the TR, sitting side by side in the couch, still dressed formally like they were in the morning. They tell me to calm down, it's only 6:30, the fireworks will not start for another two hours. They both look tired, but my mother even more so. Her eyes are a bit closed, her cheeks are flushed, and she has a strong smell of, what I later learned was, Scotch—the tea had transformed itself into an early aperitif—but she looks much more relaxed than she did in the morning. I decide to lie down on the other couch and watch the TR with them. We are watching a documentary about the war, how it was fueled by a few actors who worked adamantly to divide local populations, blaming the poor and the immigrants as is nearly always the case, purely for self-gain.

Both my brother and my sister come back a little after I did and join us in the living room. Both look tired too. The stream finishes at 7:30. My parents then arrange a little potluck by taking out all the leftovers from the refrigerator and placing them on the table. We end up with close to ten small plates that we all share, while talking about our day and the documentary. They all congratulated me for winning the bike race and for being in one of the winning groups for the world peace game.

We then all get ready to leave for the fireworks. My father tells me to pick up a sweater since it is getting chilly. We then decide to walk together to the main square. Once we arrive, I see some friends further away, in the distance, running around. They seem to be having fun, but I prefer to stay with my family. My brother teases me a bit, as usual, and while I say I hate it, I keep going back to him and seek his attention. My sister plays with us as well.

At 9pm sharp, all city streetlights are dimmed, the fountain stops running, everyone becomes quiet, and all heads turn towards the same direction. The sky is perfect for the fireworks; no clouds, no wind, just a starry night waiting to be lit. Then, the fireworks begin. One after the other, ephemeral shapes of intense color and immense beauty appear in the sky. The thing about fireworks is that they are by nature impermanent. They come to life and disappear in a matter of seconds, leaving us little time to appreciate them. Therefore, they force us to forget everything, to forget our personal troubles, our differences, and simply appreciate what we see appear before us. This is why they have become a symbol of world peace. While they are ephemeral and while they are created by explosives, fireworks bring us together, they force us to focus, enjoy the present time, and they leave us with an impression of awe. There couldn't be a better symbol of peace!

At the end of the show, completely autonomously, all the infrastructure that was turned down or off for the fireworks comes back to life. The city streetlights light up. Otos start moving around. People's devs start to act up. I can even hear the solid waste generated today being sent for processing and recovery. People are initially quiet, still dazed after the fireworks display, and slowly, voices start rising and laughs can be heard.

My brother leaves us to go see friends, and my parents, sister, and I turn around and walk back home. Once I am home, I drink some water, brush my teeth, and lie down in my bed. In a matter of seconds, my body plunges back into the cosmos. What a great Peace Day.

Biography: *Professor Sybil Derrible is a Professor of Sustainable Infrastructure in the Department of Civil, Materials, and Environmental Engineering at the University of Illinois at Chicago (UIC). He is also the Director of the Complex and Sustainable Urban Networks (CSUN) Laboratory. Prof. Derrible is originally from the small French archipelago of St Pierre and Miquelon that has a population of 6,000 people. After finishing high school, he received degrees from Imperial College London in the United Kingdom, from the Ecole Centrale of Lyon in France, and from the University of Toronto in Canada. He also worked for a year in Singapore before joining UIC in 2012 and he was a Visiting Professor at the University of Transport Technology (UTT) in Hanoi (Vietnam) in 2019. Prof. Derrible is passionate about redesigning cities to make them smart, sustainable, and resilient. He has traveled all across North America, Europe, and Far East and South East Asia, so make sure to ask him what his favorite city is!*

Acknowledgements: The author would like to thank Marie-Agathe Simonetti and Eugene Mohareb for providing comments and editing the story. The author would also like to thank MyLife Coffee Shop in Ho Chi Minh City (Vietnam), where most of his reflection was written, for providing tasty cà phê sa dá.4

— The Future of Today's Infrastructure —

Matthew Eckelman[†]

Abandoned highway, Milton, MA

[†]Northeastern University | m.eckelman@northeastern.edu

This photograph shows a closed segment of the old Rt. 128, which for decades served as part of the 'Circumferential Highway' around Boston, MA. This segment is slowly being reclaimed by nature, completely unsigned and unknown to passersby driving on the new interstate highway that runs nearby. It is now part of the Blue Hills Reservation, a state park, and is physically disconnected from any other paved roads. Walking (or biking) along its length, one cannot help but wonder how many of today's roads will look like this in 2100.

In a future of new technology, materials, and modes, what will become of our current infrastructure that is no longer needed? Will we simply build our new infrastructure on top of what we have, perpetuating the land use decisions and ecological impacts of past generations? Will we adapt these spaces and use valuable rights of way to tie our communities together in new ways, as we have converted old canals and railroad beds to recreation and commuting paths? Will we abandon it to time, as with the highway above and so many obsolete pipes and piles left in the ground? While visions of future cities often center on marvelous new creations, the greatest opportunities in the future of infrastructure may be in cleverly repurposing of what we have in the present.

Biography: *Matthew Eckelman is an Associate Professor of Civil & Environmental Engineering at Northeastern University in Boston, MA. His research and courses focus on urban infrastructure, sustainability, and energy/emissions modeling.*

— How reliable is the green in green infrastructure? —

Peter M. Groffman[+]

The capacity of natural ecological systems and processes to contribute to urban function, resilience and sustainability has received increased attention in recent years. Much of this increase has been driven by the emergence of "green infrastructure" to treat urban stormwater and provide multiple ecosystem services (Tzoulas et al. 2007, Green et al. 2016) and recognition of the ability of natural ecosystems to recover from extreme disturbance (IPCC 2012). Yet there are many challenges for using ecological concepts and specific aspects of ecosystem structure and function to support green infrastructure in cities. Most of these challenges arise from the variable and dynamic aspects nature of ecological systems, which are often not well understood or recognized by engineering and planning communities.

Of primary concern is the reliability and capacity of ecological processes related to the dynamics of water, energy and nutrients in the environment. There is a strong need for information on ecological processes to address questions such as how much water can be absorbed in green infrastructure features and how this might be affected by vegetation composition, which in turn is affected by invasive species and diseases, human preferences and management and a changing climate. For example, while we have seen that sand dunes have an impressive capacity to buffer storm surges, how will the interplay of species, climate and management affect the reliability of this function over the long term, i.e., will they be more or less reliable than a giant wall? While disciplinary understanding of these processes in many ecosystems is deep and wide, there is a clear need for synthesis and interpretation of this knowledge for urban resilience and green infrastructure efforts.

Thinking about stability, resistance and resilience in ecological systems has a long history, dating back to the 1960s (Odum 1969, Holling 1973, Bormann

[+]City University of New York and Cary Institute of Ecosystem Studies | pgroffman@gc.cuny.edu

and Likens 1979). While terminology has varied, much of the analysis has focused on how much disturbance systems can absorb and maintain their structure and function, e.g., how hard does the wind have to blow before a tree blows over and how quickly they recover from disturbance, e.g., how long does it take the forest to regrow and establish control over water and nutrient dynamics after the trees fall over. The idea of "ecological thresholds" emerged in the 1970's suggesting that ecosystems can have multiple "stable" states, depending on environmental conditions (Holling 1973, Beisner et al. 2003), e.g., shifts from clear to turbid waters or from grassland to shrubland (May 1977, Scheffer et al. 2001). A major theme that has emerged in basic science ecology over the past 20 years has been "disequilibrium," variable states and "a flux of nature" are much more the norm for ecosystems than more constant and predictable equilibrium and "balance of nature" conditions (Wu and Loucks 1995).

There is a strong need to translate ideas and research about ecological stability, resilience and thresholds into practical management of actual environments (Groffman et al. 2006). If we wish to use plant, soil and microbial processes to achieve specific functions in green infrastructure systems, we need some sense of how likely the green elements of that green infrastructure are likely to stay green. There is a great hope that these elements can be a fundamental contributor to the resilience of urban infrastructure, but there is a strong need for analysis of if this is true and for communication of these issues to engineering and planning communities involved in these efforts.

Incorporation of ecological concepts of resistance, resilience, thresholds, and alternative states into the design and management of green infrastructure should result in real improvements in green infrastructure over the next decades. Stronger integration between ecological, social and technological disciplines that results in better understanding of the reliability of the ecological components should result in green infrastructure in 2100 that is more reliable, for more functions, in more places.

Biography: Peter M. Groffman is a Professor at the City University of New York Advanced Science Research Center and Brooklyn College Department of Earth and Environmental Sciences, and a Senior Research Fellow at the Cary Institute of Ecosystem Studies. He has research interests in ecosystem, soil, landscape and microbial ecology, with a focus on carbon and nitrogen dynamics.

— Fortress infrastructure and the future of urban policy: a playful prediction —

Kris Hartley[†]

The future had been quite a disappointment. If anything evolved more quickly than technology, it was expectations. Aspirational visions of flying cars graced magazine covers in the 1950s, but by the 2020s materialized into scarcely more than delivery drones and clumsy experiments with human jetpacks. The virtualization of life was a welcome convenience but also a social and political peril. Unshakeable faith in exponential progress still led many to believe that by 2100, society would be an unrecognizable mash-up of cyborg robot humans and cities hovering majestically over flooded continents – among other sci-fi reveries. The truth was far less sexy, and the collapse of rationalism a postmodern mess.

By the mid-21st century, industrial innovation was spurred, as it had often been, by government investment in repressive infrastructure for surveillance and militarized policing. The work of the best minds was applied to protecting private assets and advancing associated policy agendas. Anti-technology and anti-science populism, embraced by both the political left and right, represented the only substantial challenge to this regime. The besieged corporate and political elite doubled-down on technology and repression as they hemorrhaged legitimacy during a neo-Luddite snapback. America's religiously inspired Millennium Dispensation Party abandoned the ethno-populism it rode to power; in its authoritarian dominance, it promoted a political narrative of salvation from perpetual crises in public health, the economy, and climate. If there were any existential rescue available, however, it was not for the masses but from them.

The promises of perpetual technological advancement stalled at the 4th industrial revolution. Innovators had been flirting with ideas like neuro-hacking before political pressure sent them into the black market, where poor coordination and tribal sabotage destroyed creative momentum over a century in the making. Meanwhile, lived experience muddled along in ways that seemed to

[†]The Education University of Hong Kong | hartley@u.nus.edu

the masses largely unchanged, save the worsening air quality, migration from in-
undated coastal areas, and withering job opportunities. Leaders begrudgingly
accepted universal basic income programs and (barely hiding their laughter)
publicly spun social welfare as pushback against elite subjugation.

Serving the preservation of technology's political legitimacy, apocalyptic
narratives about gathering ecological crises aimed to boost popular demand for
'smarter' and more protective infrastructure. Blunt interventions included the
seawalls of Venice and Tower-of-Babel-scale projects like atmospheric chemi-
cal seeding to block solar radiation and purify noxious urban air. These initiatives were the apotheosis of instrumental rationalism, as natural phenomena
were treated as crude affronts to humanly logic. Engineering solutions to problems caused by collectively irrational human behavior missed the point entirely.
Technology swept the byproducts of wasteful and indifferent human behavior under the rug, saving society just well enough to avert systemic collapse.
However, the bulge under the rug grew harder to ignore, as technology bore
an increasingly burdensome load. Indeed, the public champions of paternalis-
tic technocracy had long acknowledged its imminent futility – albeit in private
company.

The masses were disciplined over time less by religion than by fear of crisis and last-ditch hopes in technocratic fundamentalism. They were also disciplined by the controversial 'outworlding' initiatives in which humanity made its
first settlement venture into outer space with Mooncatraz and Marcatraz. Progressive 'reworlding' movements for exiled prisoners were summarily squelched.
In acts of desperation, some people committed crimes in hopes of being out-
worlded, as total captivity anywhere else was preferable to pseudo-freedom on
an increasingly repressive and decaying Earth.

After the effective merging of private and public sector apparati, the global
alliance of techno-corporate-governments (think: 'E-Parliament, brought to you
by Google') was also developing a clandestine exit venture of a different sort –
escape and exclusion to serve a panicky elite. A fortress-style infrastructure
would create a safe haven for those who de-sovereigned their wealth by relin-
quishing country-based citizenship in exchange for the UN's Platinum Global
Nationality scheme. The result was a global-scale neo-Feudalism in which elites
side-stepped the burdens of local connection, producer accountability, and tax
responsibility. Fortress infrastructure also provided retreat from the consequences of resource degradation, pandemics, and the rage of exploited masses.
In what became a once inconceivable manifestation of Henri Lefebvre's concept
of planetary urbanization, 'urban' as an abstract concept morphed into 'galactic'

– describing elite off-world realms served by the resources and efforts of those left behind. Earth, once the totality of the human realm, was in the early stages of becoming a clientelist hinterland.

In an existential sense, the reliance on ever more sophisticated infrastructures to address wicked problems like climate crises exposed a profound flaw in human logic – the idea that rational interventions, done repeatedly and forcefully enough, could solve problems having complex and irrational roots; that chaos could be measured, tamed, and normalized. This idea may be familiar to anybody who has kicked a faulty appliance (an approach that works surprisingly often and can be a cathartic release). Instrumental rationalism came to embody the fallacy that well-designed and well-executed policies excuse society from uncomfortable reflection about the pathologies of selfish behavior.

Technological progress allowed the measurement of problems and tightly calibrated application of solutions to perpetuate a doubling-down on technocratic 'fixing' – in particular, the idea that policy instruments rightly designed and calibrated could offset the unsustainable behaviors of individuals and societies. According to Hartley et al. (2019), "technocratization and the data-driven movement are perilously enamored with empiricism as their legacy, reductionism as their problem-framing approach, and initiatives like smart cities as their prescriptions; however, they offer at best an incomplete view of the factors that converge to generate existential crises" (p. 177). The doubling-down on technocracy throughout the 21st century further isolated major policy decisions to the elite halls of power, privilege, and education – paradoxically, technocracy perpetuated elitism by defining the terms truth while, in its declared objectivity, advertised itself as a means for overcoming the same elitism.

At the same time, thirst for democracy and civic participation lingered in what elites disparaged as an anachronistic logic. Alternatives to instrumental rationalism and technocratic feudalism seemed far beyond the realm of common-sense as the apostles of technocracy defined it. Nevertheless, where there was political purchase in reviving democratic ideals, some movements avoided marginalization or erasure; their prominence could be compared to that of fringe religions. The two principal alternatives were, like the Green Party of the 20th century, loosely networked global communities: the Enchanted Wisdom (EW) movement and the Neo-Know-Nothings (NKN; named after an American xenophobic political party in the 19th century).

Some argued that the reaction to rigid technocracy – and the splintering of that reaction into two revivalist movements – had roots in the constructivist and post-modernist movements of 20th century academia, which critically assailed social and political systems to make legible the sins of elite capture. Regardless of its true epistemological origins, the late 21st century's deconstruction of power aimed to make ideological space for alternative narratives; by 2100, the void had been filled by political opportunism and the revival of near-dormant ideologies. The profound cyclicality was not lost on reflective observers.

The EW movement sought to elevate the profile of folk, local, and indigenous knowledge that in many respects linked the empirical with the spiritual and mystic. The movement was the result of an unlikely alliance between Western religious groups and the progressive political left. While these odd bedfellows found little agreement on practical policy issues, the effort to supplement or supplant technocracy as a policymaking epistemic bound them and produced a movement whose size and voice were difficult to ignore.

The moment was welcoming to such a movement. Normalization of modern life and the march of ostensibly apolitical rationality across geographies had not completely muffled cultural markers and expressions. The world maintained diversity and place-based distinctiveness, with the broad message of folk empowerment finding currency in assertions of local identity. Rather than being an all-or-nothing prescription – the method by which 20th century economic and government reform movements had been imposed – the EW movement required little ideological disciplining and instead valorized a broad range of indigenous thinking, being, and doing that defined human societies for millennia.

The NKN movement sprang from similar political grievances but diverged starkly from EW. With ideological origins in a small global network of nativist authoritarians and their political and commercial enablers, a secular right emerged in which ethno-culture wars were waged but without an overtly religious flavor (the constraints of dogma proved too restrictive and it was becoming more difficult to reconcile them with the personal behavior of the movement's leaders). NKN gathered momentum under the allure of drain-the-swamp-style populism, with calls for power deconstruction carefully crafted not to threaten the movement's illiberal political beneficiaries. NKN solidarity emerged paradoxically around anti-elitism, offering aggrieved parties a chance to indict the hobgoblins of conservatism across all forms of political, cultural, and scientific progressivism.

If the EW movement lacked a single coordinating ideology, NKN was messier still; it had much to oppose but few novel ideas to fill the void. Dusty, unread copies of Atlas Shrugged and the New Testament still commanded privileged space on coffee tables, as virtue signaling never lost its potency. The movement's power structures were buttressed by cultish appeals to personality and mythical individual greatness not unlike those deployed to protect the legitimacy of 20th century authoritarians. A chauvinist shoot-from-the-hip tenor and Wyatt Earp frontiersmanship underscored not only the movement's ideological emptiness but also the distressing reality that popular support could still be solicited through age-old folksy, individualistic, and anti-intellectual rhetoric. The ability and willingness to resist poorly veiled agitprop was an individual virtue that evolution seemed to make no progress conferring.

For a brief moment during the apex of 'modernity' – the mid-20th century until the early 21st – the prospect of saving humankind from the collective crisis of innate individualism appeared to be achievable and generated considerable chatter, from multi-lateral summits to academic journals. Nevertheless, antiseptic solutions borne of epistemic rationalism and their enabling infrastructures and technologies were thin gruel for sobering political realities. The howls of techno-state evangelism were not enough to temper popular agitation. That society's ability to evolve out of existential crises depended more on better thinking than on better technology was a nuance lost to the opportunistic politics of absurdity, anger, and willfully obtuse literalism. Technology would go only as far as human bias allowed, and that was, predictably, not far. Ultimately, there was still no way to redeem the greatest failure of organized society – the inability to reconcile individual interest with collective welfare. Society's flaws were, as always, aggregations of personal ones.

The vision of shepherds wandering the ruins of the Roman Forum centuries after the collapse of one of history's most advanced civilizations, as portrayed by painters Jacob de Heusch, Giovanni Paolo Panini, and others, profoundly reflects the circularity of societal evolution. The infrastructure of 2100 appeared to its contemporary observers as rational and benevolently aspirational as the Roman Forum must have in the era of Constantine. A return to agrarian anarchy – the Hobbesian 'natural condition' that social organization and cooperation were meant to rationalize and pacify – must have seemed inconceivable. However, the rosy imaginaries of 21st century progress outpaced human abilities, as surely as the swallows to Capistrano each spring. In 2100, as in Constantine's Rome, it must have at least been comforting to think that this time would be any different.

Biography: *Kris Hartley is an Assistant Professor in the Department of Asian and Policy Studies at the Education University of Hong Kong. He researches public policy and administration with a focus on environment and technology. Kris is also a Nonresident Fellow for Global Cities at the Chicago Council on Global Affairs and an Affiliated Scholar at the Center for Government Competitiveness at Seoul National University. In 2020 he served as a Fulbright Scholar at the School of Public Policy at Chiang Mai University in Thailand. Kris's research and consulting projects are connected by the overarching theme of new public policy models for the 21st century. He holds a Ph.D. in Public Policy from the National University of Singapore and a Master of City Planning from the University of California, Berkeley.*

Reference

Hartley, K., Kuecker, G., & Woo, J. J. (2019). Practicing public policy in an age of disruption. *Policy Design and Practice*, 2(2), 163-181.

— More than 35 million people in Equinox Celebrations —

Daniel Hoornweg[+]

Dakar, Mar 20, 2100

World's largest Equinox celebrations kick-off Saturday, March 20, 2100 in Dakar, Senegal.

[+]Ontario Tech University | daniel.hoornweg@uoit.ca

Mayor-Ambassador Diouf sounded the air horn at 1:05 pm today to officially launch celebrations and the start of the vernal equinox parade.

In a well-crafted speech delivered to an enthusiastic audience Maria Diouf reminded the crowd:

"Seventy-five years ago in La Paz, Bolivia the cities of the world started to march together. They came together, to fight together pandemics, climate change, inequality, ecosystem and biodiversity loss, and our baser instincts to want to divide ourselves. Today marks the 150th Equinox parade and on behalf of the people of regional Dakar, we welcome the world. I am happy to be sharing today with Mayor-Ambassador Chal la Flamme of Paris (last Equinox host) and Mayor-Ambassador Beatriz Miranda of Belo Horizonte (the next Equinox host). We also welcome representatives of more than 400 cities, and we recognize and congratulate Mayor-Ambassador Suhadi of Jakarta, the winner of most improved sustainable city this year.

Ms. Diouf went on to highlight the history of Cities and Sustainability. "Our cities are as strong as the bridges we build between our communities, with other cities, with our rural and agriculture partners, and with our national governments. Our strength is that 75 years ago, cities came together, to collaborate with each other, to work with all communities across our urban regions, and to speak with a common language. Cities agreed, 75 years ago, to talk with each other through common measurements and understanding. One city, one standard, one world. One amazing success story - Let's continue to walk together."

The horn sounded, the mayor-ambassadors joined hands and led the procession through five kilometers of Dakar's decorated streets. This year's event included more than 500 musical bands and some 500 floats and dance teams. The global viewing audience is believed to have surpassed last year's record, with more than 8 billion people watching around the world.

The equinox festivities are accompanied with the bi-annual release of the common city sustainability indicators. Regional-Jakarta showed the most improvement among the 508 participating urban areas with populations over one million. The city-sustainability.com info-site received more than 1 billion visits in the last two weeks as people around the world provided input for local sustainability progress.

With updates to the sustainability ratings of the world's cities, the 'factor of sustainability' rating for all existing and proposed infrastructure projects serving these urban areas was automatically re-calculated (about 75,000 projects with total financial values above $100 million each). More than 800 projects shifted to 'highly sustainable' ratings, while 250 projects moved out of the 'highly sustainable' category. This year also saw a record $1 trillion available as preferred funding for the approximately 5,000 infrastructure projects evaluated as 'highly sustainable'.

The World Federation of Engineering Organizations and the International Federation of Accountants issued their joint statement, confirming the infrastructure ratings and awarding their bi-annual Top Ten sustainable infrastructure projects. Top projects included: organic waste digester, Kolkata; household product delivery, Beijing; fast-travel, Toronto-Montreal; water supply, Mexico City; uranium-thorium batteries, Stockholm.

Dakar's bi-annual Equinox conference hosted a record 430 mayor-ambassadors, 88 heads-of-state and more than 10,000 government officials. Once again, meeting the deadline of today's parade, agreements were finalized in three areas: data security protocols post Tetra-Tynes corporate breech; protection of song-bird and cetacean migratory routes; fiscal support to New York City post Hurricane Donald.

The cities (urban areas) are represented by mayor-ambassadors selected by their communities for this once-in-a-lifetime honorary role. Mayor-Ambassador Diouf was selected for her 23-years of service in Dakar City Council and strong ties to neighboring Thies (where she now resides).

This 75th year anniversary of 'cities and sustainability' recognizes the city innovation platform developed and managed by cities and their countries, as well as 75 years of continuous monitoring of energy and material flow in cities – and related sustainable development goals. The 'cities and sustainability' platform was awarded the Nobel Peace Prize in 2035 when the independent evaluation showed that the program substantively shifts infrastructure financing toward greater sustainability (this year's review showed that this has resulted in a 20 percent reduction in global military spending).

Other Equinox celebrations around the world included a Still-Rolling Stones concert in London, an orchestra lightshow in Sydney, flying of the dragon-drones in Montreal, and the traditional Equinox football match between Rio and Sao Paulo (this year in Sao Paulo). For 75 years, on the vernal and autumnal equinoxes, the world's cities come together, represented by regional mayor-ambassadors, take stock of their efforts toward sustainability, re-evaluate their urban infrastructure, push their national and local governments to strive for genuine sustainability, and as much as possible, take a break, and have a good party.

Reported by Daniel Hoornweg, Richard Marceau Chair, Ontario Tech University, Canada.

Biography: After almost 10 years in local government and 20 years in the urban sector of the World Bank, Dan returned to Ontario to obtain a PhD in Civil Engineering (Sustainable Cities) at the University of Toronto (2015) and lead energy and urban system research at Ontario Tech University as the Richard Marceau Chair. Dan explores energy and material flows of cities, and has been working on climate change since 348 ppm.

— "Only you know if we did it" —

Nadine Ibrahim[†]

It's August 18, 2100, just north east of Reykjavik, Iceland, and I'm surrounded by such picturesque glaciers – fabulous and very scenic. I'm gathered among a group that walked up to visit the site of the volcano where the first ceremony to commemorate the loss of the glacier Okjökull was held on August 18, 2019. The glacier had lost its glacier status and has become "Ok" since then, losing the -jökull (glacier in Icelandic) part of its name. Today marks the 81st anniversary of this ceremony when the plaque was installed, and it reads as follows,

> "A letter to the future. Ok is the first Icelandic glacier to lose its status as a glacier. In the next 200 years all our glaciers are expected to follow the same path. This monument is to acknowledge that we know what is happening and what needs to be done. Only you know if we did it. [Signed] August 2019. 415ppm CO_2"

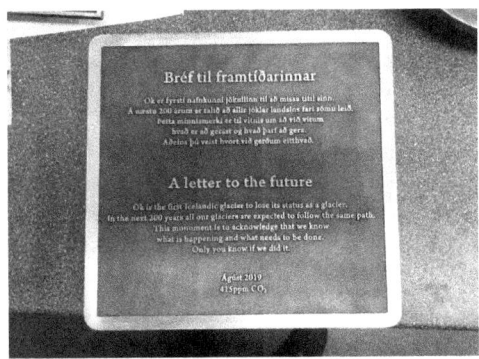

Image Reference: Okjökull glacier commemorative plaque by Rice University, Asav is licensed under CC BY-SA 4.0 on Wikimedia Commons

In August 2019, I was teaching Engineering and Sustainable Development to the second year civil engineering class at the University of Waterloo, and as

[†]University of Waterloo | nadine.ibrahim@uwaterloo.ca

I wrapped up the last lecture and did an exam review, I ended my slide presentation with an image of this copper memorial plaque. The class went silent, and I left it on that note, with the future in their hands to turn this around! As the year came to a close in 2019, people around the world knew then that the coming decade was going to be the golden age of infrastructure. It was the infrastructure that holds us together that was going to be the solution, and sustainable infrastructure was going to be our only way out of this climate change mess.

In 2100, it has become not just a decade of transformational change, but many many decades, and we dedicated all our efforts to solve the climate crisis. We're still not even halfway towards the 200-year prediction engraved on the plaque, and as we look around, we revel with admiration at those who came before us and the science and evidence that they shared with the world back in 2019 to urge them to take action. It was mostly the younger generation that pushed the others to make such bold decisions in renewable energy, building science, transit, water conservation, waste disposal, forest management, among many others.

The group I am with in Reykjavik is getting ready to see each other again when we travel to Lagos, Nigeria to attend the UNFCCC Conference of the Parties, COP106 this year. The world's most populated city is indeed Lagos. Today, in 2100, the world's most populated cities are mostly African. Actually 5 out of the top 10, or 13 out of the top 20 most populated cities are in Africa, as predicted by Hoornweg and Pope* many decades ago in 2017. The COP has become the largest and most attended event of the year, thanks to its growth into a very influential meeting of the minds, and global leaders now look to align their upcoming political (and election) agendas based on what the COP deems to be the most important issues of the year.

I had predicted in my PhD thesis back in 2015 that cities will likely be able to meet and surpass their 2020 climate targets, but doing more of the same will not be enough to get them to their 2050 climate targets without transformational change. Climate targets then allowed us to limit the concentration of CO_2 in the atmosphere to 450 ppm (parts per million), which scientists had later decided that it was a very irresponsible strategy with dangerous future consequences. To even dream of having a 50:50 chance of limiting global warming to 1.5°C, the global economy had to become carbon neutral by 2050. The 1.5°C pathway lead us to net zero emissions by mid-century, and we have been devoted to guarding that progress ever since. 350 ppm was considered a safe level, and less than that would be that much closer to pre-industrial concentration.

Today, we stand in awe of the glaciers around us, and marvel at the plaque, and think to ourselves, indeed, we did it!!

Signed: August 18, 2100. 280 ppm.

We left fossil fuels in the ground.
We built cities for the convenience of people, not the comfort of the cars.
We built taller buildings and avoided urban sprawl.
We lived in buildings that are net-zero.
We preferred commuting by transit, and left the cars behind.
We developed an extensive network of high speed rail.
We reverted to what we have always done – walking!
We used the sun, wind and geothermal energy to the best of human potential.
We cut out red meats from our diets.
We planted more trees and prevented deforestation.
We created innovations to capture carbon.
We consumed less and reused more.
We recycled everything.
We created a circular economy, and we became one with nature.
We formed the kind of cities that we've always
dreamed of living in. Finally![‡]

[‡]Hoornweg, D., and Pope, K., 2017. "Population predictions for the world's largest cities in the 21st century." *Environment & Urbanization*, 29(1): 195-216.

Biography: *Nadine Ibrahim is a Lecturer in the Department of Civil and Environmental Engineering and holds the Turkstra Chair in Urban Engineering at the University of Waterloo. She comes from a cross-section of industry and academia in the areas of urban infrastructure, sustainable cities, and sustainable development. She holds a BASc, MASc, and PhD in Civil Engineering, and a Certificate of Preventive Engineering and Social Development, from the University of Toronto. Her work in cities has taken her to the Middle East and North Africa region where she worked in international development, and to cities in Southern Ontario, Canada where she worked on asset management, and to a dozen African cities during her post-doctoral research. She is passionate about engineering education and chairs a special interest group at the Canadian Engineering Education Association called the "Engineer of 2050," defining future skills for engineering education. Her research looks into urban engineering, sustainable infrastructure, material and energy flows, and climate action in global cities, megacities, and mega-regions. Nadine teaches courses in civil engineering systems, decision-making, sustainable development, sustainable cities, and urban prosperity. She enjoys very much teaching undergraduate engineering students how to calculate parts per million (ppm) so that they can interpret CO_2 concentrations in the atmosphere – to her, that is a key part of sustainability literacy!*

— The present tense of long-term thinking —

David M Iwaniec[*]

As part of the Urban Resilience to Extreme Event project, researchers and practitioners collaborate to explore current vulnerabilities, anticipate future trends, and envision positive futures for their cities in Latin and North America. In each of these cities, the researchers and practitioners alike, have at some point during the project confronted fundamental tensions on the needs and roles in doing long-term, anticipatory thinking:

- Why imagine 2100 when there are such pressing and immediate needs today?

- Why anticipate infrastructure futures when faced with a rapidly changing world of disruptive technologies, unpredictable surprises, and highly uncertain outcomes?

- How salient and relevant to decision making are visions of infrastructure in 2100?

We do need to develop solutions to address urgent, immediate challenges. We also need to develop solutions that might require longer time horizons to unfold. An essential tension of anticipatory thinking is how to make this coherent. That is, using long-term futures to make better decisions today.

The UREx project pursues this by conducting scenario co-development workshops in each city to (1.) explore implications of existing goals, initiatives, and targets and (2.) envision desirable-plausible futures of sustainability, resilience, and equity.

For this first objective, extrapolative approaches are used to explore long-term implications of existing, formalized goals and targets (e.g., from municipal, NGO, and community planning and governance documents; Iwaniec et al., in press). That is, participants, evaluate current goals for their city through the

[*]Georgia State University | diwaniec@gsu.edu

103

lens of whether they sufficiently address persistent and emergent challenges. Here, long-term futures are used to explore if existing goals are transformative enough and the potential consequences of current trajectories.

The second objective focuses on the capacity to articulate positive futures —desirable pathways composed of multiple intragenerational and intergenerational values (Iwaniec et al., 2020). Developing scenarios for long time horizons allow participants to look past barriers (inertia, lock-in, balancing feedbacks) to changing the status quo. That is, envisioning a distant future that is not constrained by current governance structures or existing infrastructures (i.e., the way things work now). Here, long-term futures are used as an instrument of innovation and creativity—an acknowledgment that we have agency to shape our future. Barriers to change are reframed as the 'places to intervene' and as opportunities to reimagine infrastructure.

While much of the problems that our communities face are products of past solutions, it is untenable to suggest that long-term thinking could ever overcome this cycle when developing new solutions. Rather, the goal is to build the anticipatory capacity of our communities: to imagine possible consequences, negotiate tensions and trade offs, and discover opportunities in uncertainty or in what might otherwise be considered inevitable. A privilege that should not be limited to just a few.

Biography: David M. Iwaniec is an Assistant Professor of Urban Sustainability at the Urban Studies Institute, Andrew Young School of Policy Studies at Georgia State University. He is a sustainability scientist researching anticipatory and systems approaches to advance urban sustainability, resilience, and justice. He conducts research to learn from, and improve, the governance of urban social-ecological-technological systems. His research focuses on the co-development of scenarios and transition pathways for positive futures of urban transformation.

— As we don't want to burn fossil fuels, we don't need to transport them nor build the infrastructure to do so —

Chris Kennedy[+]

Back in 2011, I took a leave from my position as an engineering professor at the University of Toronto and headed to Paris for a year to work at the Organization for Economic Cooperation and Development (OECD). It was fun living in Paris; we rented an apartment in the 3rd arrondissement, surrounded by patisseries, and my kids attended the local school *École Saint-Martin* (Fig. 1). At OECD, I participated in a broad study of how to mobilize billions of dollars of investment in low-carbon, climate-resilient infrastructure. As part of this work, I wrestled with the question of how much infrastructure investment is globally needed to steer us towards a low carbon future, the focus of this short piece.

A key input to the question was the 2012 *Energy Technology Perspectives* report, produced by the International Energy Agency (IEA), a sister organization to the OECD. In the report, the IEA provided estimates for the future costs of power systems, industry, buildings and transportation vehicles under a low carbon future (in those days the target was under 2 deg. C) versus business as usual (3 to 6 deg. C). Overall the IEA study showed that an incremental investment of about 450 billion USD per year would be required for a low carbon trajectory, of which 300 billion USD/yr. was for buildings. I suspect that the estimate could be updated, and hopefully is lower given progress with technologies such as renewables and electric vehicles. Still, how does the world find an extra 450 billion dollars per year?

Another piece of the answer was in a report *Strategic Transport Infrastructure Needs to 2030*, published in 2012 by the OECD's International Future's Program. At first this report gave me great concerns about carbon emissions, as it described how the world was expected to undergo: a doubling of air passenger traffic in 15 years; tripling of air freight in 20 years; and quadrupling of

[+]University of Victoria | cakenned@uvic.ca

Figure 1. The dining table at our Rue Greneta apartment. The kids had eaten so much cake they had turned into zombies.

port handling of maritime containers by 2030. I didn't do the calculations, but knew that this amount of air and marine activity would translate into substantial greenhouse gas emissions – significant on a global scale.

The *Strategic Transport Infrastructure Needs* report, however, was written in a business-as-usual context – without factoring in the changing infrastructure requirements under a low-carbon future. Amongst the estimates for future infrastructure needs, I found some hope for a low-carbon world. The report projected future infrastructure investments of 245 billion USD/yr. for roads, 120 billion USD/yr. for railways, 120 billion USD/yr. for airports, 40 billion USD/yr. for ports and 155 billion USD/yr. for oil and gas distribution. Under a low-carbon future – in which the world dramatically reduces its combustion of fossil fuels – we could, however, save on some of these infrastructure costs. Surely much of the 155 billion USD/yr. on oil and gas infrastructure would not be needed. With more compact cities and more emphasis on sustainable transportation modes, perhaps we could spend less than 245 billion/yr. on roads. If carbon was appropriately priced, then maybe the huge expected increases in air traffic could be reduced, lessening the need for new investments in airports. My hunch was that some of the 680 billion USD/yr. on transport infrastructure

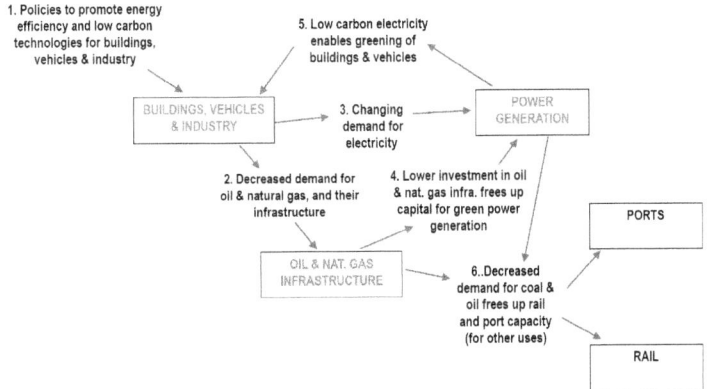

Figure 2. Key financial and physical interactions between infrastructure sectors contributing to low-carbon economic growth (simplified version of Figure 8 in Kennedy and Corfee-Morlot, 2013, Past performance and future needs for low carbon climate resilient infrastructure–An investment perspective. *Energy Policy* 59, 773-783).

could be directed towards the incremental costs of low-carbon infrastructure.

The future needs for rail infrastructure in a low-carbon world were particularly intriguing. Maybe there would be increasing demands for passenger rail to offset travel by road or plane. One the other hand, a large amount of the freight carried on railways is coal! I dug up some numbers that surprised me. Coal accounts for close to 60% of freight-tonnage on Chinese railways. Even in the United States in 2007, coal was 44% of the freight-tonnage carried. In subsequent work I took a closer look at China's *Medium to Long-term Railway Network Plan*. I realized that much of it would not be needed if China pursued its *2050 High Renewable Energy Penetration Roadmap*, which would reduce coal use by 75%. The big lesson that emerged from my work in Paris was that as we don't want to burn fossil fuels, we don't need to transport them, and hence can save billions on the costs of infrastructure.

Of course, it's all well and good to suggest that we can pay for low-carbon infrastructure by not spending on oil and gas pipelines, railways and other infrastructure for transporting fossil fuels, but how could this come about. I had an ah–ha moment one evening early in 2012, at our Rue Greneta apartment in Paris. Sitting at the dining table, working through the IEA and OECD infrastructure cost estimates, I sketched a diagram showing how the infrastructure sectors were inter-related (Fig. 2). By starting with policies that promoted energy efficiency and low-carbon technologies for buildings, vehicles and industry it was possible to kick-start a low carbon virtuous cycle. I joked to my wife, *"Hey, I have a plan to save the world!"*

Biography: Chris Kennedy is Chair of Canada's Green Civil Engineering Department at the University of Victoria. He is a former President of the International Society for Industrial Ecology, and author of The Evolution of Great World Cities: Urban Wealth and Economic Growth.

— A Timeline of Future Transport in Sydney as Revealed in Tablet Form —

David M. Levinson[†]

While digging for bitcoin in my copious Sydney backyard, I uncovered golden tablets revealing to me the near future history of transport in Sydney. For the public benefit, I have digitized the content of these tables to share with you, on the intertubes. I appended it with current-day hyperlinks to help the reader.

2019 The M4 East section of WestConnex opens and is heralded as a success when the ribbons are cut. It does not immediately collapse. It misses demand (revenue) forecasts.

2019 Metro Northwest opens. While technically successful, residents realise Metro is a mistake compared to trains for long distance travel as Sydneysiders want seats for long rides. Protests at public hearings evolve into riots as train chairs, brought to the hearing, are thrown by agitators. Future extensions of rail-based transit use double-decker trains. No more public hearings are held.

2019 Central business district (CBD) and South East Light Rail opens. Ridership is higher than forecasts and everyone wishes it had been constructed as a metro or train. People complain about long traffic signal delays along the route.

2020 NorthConnex opens and is heralded as a success when the ribbons are cut. It does not immediately collapse. It misses demand (revenue) forecasts, but is congested on Day 1. Investigation into the forecasts revealed that all documentation was destroyed in a fire.

2020 The final section of the M4 opens and commuters realise the Anzac Bridge will not have expanded capacity, so the bottleneck is for many travelers simply moved upstream. Long distance trips using the bridge do not save actual time. Officials say, just wait until the M4/M5 Connector Opens.

[†]University of Sydney | david.levinson@sydney.edu.au

2020 The M5 East Section of WestConnex opens and is heralded as a success when the ribbons are cut. It does not immediately collapse. It misses demand (revenue) forecasts.

2020 5G Wireless networks become available.

2021 All Sydney Buses are privatised under a set of franchise agreements.

2022 There is pressure on the government to bail out the private operator of WestConnex.

2023 The M4/M5 Connector opens and is heralded as a success when the ribbons are cut. It does not immediately collapse. It misses demand (revenue) forecasts.

2023 Parramatta Light Rail opens. It hits forecasts and no one is fired. There are no recriminations and no lawsuits. No one went to gaol. This was a first for a transport project. A subsequent analysis of the forecasts revealed that a stopped clock is right twice a day.

2024 City and Southwest Metro opens and is heralded as a success when the ribbons are cut. Train services on other lines are expanded with the new slots freed up by transferring the Bankstown Line to Metro Service. Customers on the Bankstown Line are not pleased with having to stand.

2025 Western Harbour Tunnel and Beaches Link open and are heralded as a success when the ribbons are cut. It does not immediately collapse. They miss demand (revenue) forecasts.

2025 Electric vehicles comprise more than 50% of new car sales in Australia. Motor fuel tax revenues are beginning to be decimated. The government announces it will implement road pricing over the next 5 years both to raise revenue and to manage traffic. The government buys WestConnex and other private toll roads back when it wants to implement comprehensive road pricing and realises charging a premium on certain roads is an inefficient use of public resources. TransUrban is given the contract to manage all Sydney roads. They do it for "free," charging only expenses with a 10% handling fee on all transactions.

2026 A new airport in Badgerys Creek opens. The traffic at the airport is disappointing. Nevertheless, real estate interests argue for closing Kingsford Smith Airport to create a valuable business district with a water view of Botany Bay.

2027 SouthConnex (F6) opens and is heralded as a success when the ribbons are cut. It does not immediately collapse. It misses demand (revenue) forecasts.

2028 WestConnex use is restricted to trucks, buses, and automated vehicles for safety reasons. A new fault line closes the tunnel for a year for retrofitting.

2028 All Sydney Buses are nationalised.

2029 Sydney West Train Line opens from Sydney to Bays District, Rozelle, Olympic Park, and Parramatta. The University of Sydney still has no station. Though initially planned as a Metro, it was decided to build it for conventional double-decker trains. Dual platforms on each side of the train speed boardings and alightings.

2030 6G Wireless Network becomes available. Nearly 40% of workers telecommute on any given day.

2030 Level 4 automated vehicles, which are approaching 30% of all cars, greatly expand capacity on the road network. Single-seat cars become standard.

2030 The City of Sydney bans private cars in the Central Business District. Other Councils follow in other Business Centres and High Streets. Car usage drops.

2031 Limited stop train connects the Western Sydney Airport with Parramatta and the Sydney CBD. It turns out most passengers want to go to other destinations. Driverless taxis do very well.

2032 In celebration of Sydney Harbour Bridge Centennial, Sydney closes down Cahill Expressway.

2034 Sydney's Kingsford Smith Airport is permanently closed and all air traffic is diverted to the new airport at Badgery Creek. Botany Bay is reconstituted as a new business district. Fitting with state policy, no schools are planned. Instead existing schools in Mascot are expanded to 14,000 students using demountables. (Research that showed educational outcomes improve with school size was subsequently retracted in 2056 after it was discovered to be funded by the demountable industry without proper disclosure in the acknowledgments).

2035 Sydney rebuilds Monorail to connect CBD with Darling Harbour.

2036 Green Square light rail line opens, connecting Zetland with the CBD.

2037 The Lachlan Canal was constructed so high-speed ferries can serve Western Sydney airport. Part of the canal uses a former rock quarry (the Quarry MacQuarie) turned into a lake. The newly formed lake is called the Loch Lachlan MacQuarry. A sea monster soon appears. This raises many questions. It is photographed many times, but never captured. The Lachlan Canal means Western Sydney is no longer landlocked.

Loch Lachlan MacQuarry Sea Monster

2039 Construction on a very-fast train line to Melbourne begins. It is called Macquarail. Despite many maglev and hyperloop proposals, it uses 80-year old Shinkansen technology. A series of planned communities along the path justified the rail line, providing low cost housing for urban commuters.

2040 Sea level rise begins to shrink the amount of land in Sydney. Discussion of constructing the Macquarie Seawall begins.

2041 Eastern suburbs railway line is extended to Bondi Beach. JJC Bradfield was reportedly pleased and getting a tan.

2042 Human-driven cars are banned from New South Wales roads.

2045 Sydney begins to reconstruct (driverless) tram network along historic right-of-way. Roads are closed to most other vehicles to give trams exclusive right-of-way. Cars can no longer drive on tracks. Trams are more popular than buses. It is called the Lach-rail network.

2046 EastConnex Tunnel connecting Green Square and Bondi Beach is opened.

2047 Due to declining crowds, professional sports are no longer played in stadia, but just in video studios. Many sports have been banned for human play due to head injuries. The last stadium is closed and torn down.

2048 Sydney bids for 2052 Olympics.

2048 Air transport conversion to biofuels is completed. Transport is now carbon zero.

2049 Public schools are closed. Education is replaced with personalized robotic instruction.

2050 Sydney population reaches 8 million.

2050 Australia becomes a Republic upon the Death of Queen Elizabeth II, the world's longest reigning monarch, richest woman, and longest-lived person. The new King Charles reportedly said "Finally" before passing away himself. The final season of the Crown is aired on Netflix. Transport for London had named three more train lines after her, including the Queenie, the Monarch Line, and the Corgi-Owner Line, after the more obvious Elizabeth and Jubilee lines had been used.

2051 The famous "One-track mind" paper is published. Neuroscientists discover the part of the brain that likes transport on two rails is activated under the presence of a specific genetic allele. A separate, rarer allele is found in monorail fans, but it is recessive and found only in economic development officials and amusement park designers. This genetic discovery shapes subsequent transport policy. Parliament bans genetic engineering for the purpose of creating rail-fans, concluding it is too costly for society. Secret clinics carry on these genetic modifications anyway.

2052 Australia Post workers are replaced with robots and drones. Mail delivery, by this point comprises mostly of packages, magnets for local electricians, pizza coupons, and real estate listings, is restored to 3 times per day.

2052 Sydney hosts 2052 Olympics at New Olympic Park - Emu Plains. The Olympics' mascot is the Common Myna. The site is connected by high speed transit to Penrith, Parramatta, CBD. This is the first Olympic games where up-lifted animals compete in selected competitions. An Emu named Kylo Ren wins the men's 400m dash. All of the losing humans are subsequently disqualified for drug use. A mountain gorilla wins all the shooting competitions.

2053 New Emu Plains University opens at the athletes village of New Olympic Park. It quickly rises to be ranked number 4 in the world. It is the first university in the world to admit uplifted animals. The state quickly extends high-speed trains to the new centre of scientific technology in the far west. Corporations locate R&D parks in the vicinity. People are surprised by the high ranking, but then it attracts many high quality students making college decisions totally based on university rankings, and so eventually merits the ranking. Subsequent investigation revealed an Excel error in the ranking formula.

2057 Sydney Metro is retrofitted to double-decker train service. The tunnels needed to be increased in height by 1 cm to do this. Construction lasts 5 years, but is only undertaken on school holidays. This was the first project to be built entirely by robots.

2060 Buildings in Sydney begin to be raised (not razed) to accommodate higher sea level.

2063 The first of a series of annual Frost Fairs are held in July on the Parramatta River, in and around Parramatta. Attempts at Geo-Engineering and the subsequent surprise eruption of a newly emerged volcano in the Pacific have compounding effects resulting in cold winters. Lack of insulation in Sydney homes means a surprising number of deaths occur. Insulation nanobots are deployed to insulate Australian ceilings and seal windows.

2069 Construction on the Macquarail, a very fast train line, to Melbourne is completed. Demand is below expectations. The Greater Melney Metropolitan Area is established by the government, and becomes one of the world great megacities.

2070 Sydney tram network is again the world's largest.

2074 WestConnex floods as the underground water table rises inexorably.

2076 Sydney's canal and gondola network is greatly expanded.

2078 WestConnex is turned into a giant underwater luge track in preparation for 2080 Olympics.

2079 All Sydney Buses are privatised.

2080 Third Sydney Olympics held. The main Olympic stadium is built on a floating island in the Pacific Ocean. Access is only by ferry and amphibious flying cars. Plans to extend the EastConnex Tunnel faltered. After the Olympics, the stadium is sailed to Morocco for the 2084 Olympics. This is the first Olympics with no human competitors.

2087 Sydney begins dismantling its new tram network, 'Flying trams for the future' the government says in the Press Conference. The business case for flying trams is released in 2231 after a Freedom of Information Act request by the Melney Age of Heraldry newspaper (filed in 2089). It turns out a basic math error that had been used to justify the decision resulted in a Benefit / Cost ratio of 1.3 instead of 0.3 as it should have been. The government apologise for the error, but blamed the opposition.

2090 After 50 years of discussion, construction begins on the Macquarwall, the Macquarie Sydney Sea Wall. Though CO_2 emissions have fallen to zero globally, warming and sea level rise continues. North Korea explodes its nuclear weapons to slow warming.

2095 Australian Space Sciences Agency (ASSA) constructs the Space Elevator, which is used to efficiently transport cargo to outer space. ASSA builds the earth's first interplanetary passenger ship named the Endeavory McEndeavorFace after a popular survey. Colonisation of Venus begins. The passengers consist of people convicted of crimes and offered the option of this space trip or the death penalty. Terraforming bots arrived at Venus prior to settling and had ensured the atmosphere was breathable. The planet is declared *Terra Nullius* despite indigenous populations being discovered.

2100 Greater Sydney population reaches 16 million as it annexes Melbourne to form Melney. A strong contingent had supported Sydbourne, but a man named Sid Bourne had trademarked the name and courts refused to permit it, saying confusion in the market would follow.

2132 Sydney Harbour Bridge celebrates its Bicentennial. All travel lanes have long been converted to bus and truck-only lanes. A planned Sydney walk across the bridge ends tragically when all 200,000 people walking across the bridge start to jump up and down in celebration. The jumps resulted in unfavourable harmonics. The Bridge is not rebuilt, but instead left as a ruin in monument to human folly and as a case study for Civil Engineering Bots.

2147 All Sydney Buses are nationalised.

2150 Sydney's population falls to 8 million as many residents migrate to Adelaide. No new transport infrastructure is constructed, except for the Tram network, which is rebuilt.

2200 Sydney's population falls to 4 million as more residents move to Perth. No new transport infrastructure is constructed. The Tram network is dismantled.

2200 Average speed on wired internet in Sydney rises to 20 Mbps. National Broadband Network (NBN) announces the network is completed.

2200 43G Wireless Networks become available. Speeds are approximately 1000 Exabytes per Second. This allows everyone to project a holographic image of themselves to everyone else globally, simultaneously. Netflix is still buffering though.

It's all true.

Biography: *David Levinson teaches at the School of Civil Engineering at the University of Sydney (official profile), where he leads TransportLab and the Transport Engineering group. Levinson has authored or edited several books, including A Political Economy of Access, Elements of Access, Spontaneous Access, The End of Traffic and the Future of Access, The Transportation Experience, and Metropolitan Transport and Land Use: Planning for Place and Plexus, as well as numerous peer reviewed articles. He is the founding editor of Transport Findings and the Journal of Transport and Land Use.*

— Versatile Infrastructure —

Matan Mayer[+]

Amager Bakke waste-to-energy plant, Copenhagen. Photo: Rasmus Hjortshoj (2019)

[+]IE University | mmayer@faculty.ie.edu

In a world that is increasingly virtual, physical infrastructure has never been more crucial or present. Data centers and fulfilment hubs are some of the prevalent and indispensable components of a global economy that did not exist just two decades ago. The climate crisis has also brought about new varieties in productive and defensive structures. From sea level rise surge barriers, through wind farms to solar fields, infrastructure is now more diverse and ubiquitous in our cities than ever before. While these infrastructure types are novel, in many cases their design tends to stick to a strictly utilitarian tradition that affords little to no interaction with their context. In a few rare instances, this infrastructure is elevated to serve also as a public and/or social amenity. This photo shows Amager Bakke, a waste-to-energy plant completed by Bjarke Ingles Group in 2019. Located in Copenhagen, the facility is designed as an urban hill in a region that has little natural topography of its own. In wintertime, the hill amasses snow and acts as a ski resort. Following this trajectory, how would infrastructure look like in 2100 if we fully embrace versatility? Could we envision new landscapes that are at once productive, defensive and recreational? The boldest form of infrastructure in this future might be an invisible one.

Biography: Matan Mayer is an Assistant Professor of Architecture at IE University in Segovia, Spain. His research and teaching focus on design implications of life cycle thinking in the built environment.

— Infrastructure Systems 2100: My Hope for 2020 —

Sue McNeil[†]

I love infrastructure systems. I find it remarkable that these infrastructure systems serve millions of people each day. They provide shelter, access to clean water, dispose of wastes, deliver light, heating and cooling at the flip of a switch, allow us to communicate with each other across time and space, and offer transportation by a wide variety of modes. They support economic growth and prosperity, education, health, and social interaction.

As a user of infrastructure systems, I love the convenience, comfort and service provided by infrastructure systems for relatively low cost. Yes, I am frustrated by congestion or a dropped cell phone call or the disruptions caused by extreme weather events or repairs, but overall, the service provided is reliable and meets my needs.

As an engineer, I love the way we engineer networks to provide service to many. I love seeing new construction, the repair of damaged infrastructure and the renewal of aging infrastructure. I love the fact that delivering infrastructure services requires the skills and input of not just engineers but also planners, social scientists, financiers and accountants.

As a tourist, I love seeing how infrastructure works in different places. I love visiting the ancient cities of the world and seeing infrastructure systems that have endured centuries. I love visiting the mega-cities and marveling at the services delivered by an interconnected network of infrastructure services. I love that our transportation infrastructure lets me escape to the mountains or a remote location but still stay connected, and enjoy my comforts (particularly, hot showers, and flushable toilets).

As a researcher, I love the opportunities to collect and use new and better data about the condition, use or performance of our infrastructure, to model the changes in our infrastructure over time or due to unexpected events, to explore the trade-offs between investing now or later, to find optimal or near optimal strategies for investing in these important systems.

[†]University of Delaware | smcneil@udel.edu

As an educator, I love to pitch to students the idea that we are stewards of our infrastructure. In effect we need to be responsible for not just planning, designing and building infrastructure systems but maintaining and operating these systems over their life cycle.

So, how do I expect our infrastructure systems to function by 2100, the end of this century? I expect that technology will be further integrated into these systems and our lives. We will be able to choose and manage our consumption of energy, our modes of transportation and our communications tools to meet our needs. Smart cities will be the norm; self-diagnosing, self-healing infrastructure will be prevalent; engineered materials will last longer; new construction and repair methods will minimize disruption; decisions will be data-driven and recognize the needs of the diverse stakeholders and users.

More importantly, what do I hope for in 2100? I hope we avoid unintended consequences. For example, modern infrastructure can be blamed for climate change – an unintended consequence of infrastructure improvement. I hope we will have addressed some of the important environmental and equity issues. This includes processes for building, renewing and using the infrastructure that do not degrade the environment; and access to reliable infrastructure services independent of income, race or location. I hope that sustainable, resilient infrastructure systems will become reality. I hope that we understand that the various infrastructure systems are connected. I hope that the public at large recognizes the need to pay for infrastructure – not just new infrastructure but the upgrading and renewal of aged infrastructure. I hope these infrastructure systems will be safe and reliable to support humanity's health and development. I hope that the next generation is able to appreciate the role infrastructure has in history. Seeing ancient Roman roads where chariot wheels have worn the rock away and drains moved wastes away. I hope that we can still enjoy the natural environment as green infrastructure, whether a manicured park or a wilderness area, plays an important role in our social fabric.

So what will be the role of the researcher and educator in 2010? Infrastructure systems are always dynamic. Therefore, there will always be something new to research. More data? New data sources? New and better sensors? New repair methods? New modes of transportation? New methods for generating and delivering energy? New communications methods?

Education will take on a new role as new specialized professions emerge to support these new infrastructure systems. Just as mechatronics and robotics have grown as disciplines within engineering, infrastructure engineers will need skills and knowledge that we have not even dreamed of.

Biography: *Sue McNeil is Professor of Civil and Environmental Engineering and Public Policy and Administration at the University of Delaware. She serves as the department chair and is also a core faculty member in the Disaster Research Center at University of Delaware. She is a former Director of the Disaster Science and Management graduate program and the Disaster Research Center. Her research and teaching interests focus on transportation asset management. Her most recent research includes the impact of natural hazards and climate change on physical infrastructure and asset management with emphasis on resilience.*

— Crumbling —

Martin V. Melosi[†]

Crumbling.
Crumbling under our feet.
Crumbling before our eyes.
 America's Infrastructure Report Card, 2017: D+.
Drinking water: D; Wastewater: D+;
Hazardous Waste: D+; Solid Waste: C+;
Dams and Levees: D; Bridges and Ports: C+
Inland Waterways: D+;
Parks and Recreation: D+; Schools: D+;
Energy: D+;
Roads: D; Transit: D-; Aviation: D;
Rail: B!
 Hardly grades Mom and Dad would praise—
No car for a month!

What to do?
Apparently, nothing.
Lots of talk, lots of bluster.
And corona doesn't help.
 Money for anything but…until it's too late.
Flint.
Dystopia?
That's not smart.
Is smart technology smart?
Maybe.
By 2100 will we get wise,
And stay smart?

[†]University of Houston | mmelosi@uh.edu

Biography: Martin V. Melosi is Cullen Professor Emeritus of History and Founding Director of the Center for Public History at the University of Houston. He is the author or editor of twenty books, including The Sanitary City. His latest book, Fresh Kills: A History of Consuming and Discarding in New York City, was published in January 2020.

— The Tolerant City —

Eugene Mohareb[+]

Our city is better attuned
Than the one I knew in my youth
To the rhythms of the seasons
And blending the old with the new
 Our city cools us in summer's heat
With ample shade beneath its trees
Grey makes way for lush greenery
With room for bees and birds and breeze
 Our city is much more active
Engines replaced by cranks and feet
Smog and smoke have all dispersed
Now children's laughter rises from the streets
 Our city produces no waste
Only using things that endure
And the things we cannot reuse
Are absorbed by plants and the earth
 Our city is always changing
Spaces can shift from use to use
Shielding our vulnerabilities
And serving functions that we choose

Our city is soft and porous
Not just to rain from stronger storms
Accessible to all people
Broadening out its social norms

Cities need care and renewal
So our work will never be done
This city meets all of our needs
And belongs to everyone

[+]University of Reading | e.mohareb@reading.ac.uk

Biography: Eugene Mohareb is Lecturer in Sustainable Urban Systems at the University of Reading. His research focuses on greenhouse gas mitigation from cities, including work on low carbon technology adoption and food system interventions.

— The Future of Infrastructure in an Era of Changing Complexity —

Johanna Nalau[†]

The future of infrastructure and how we have designed our cities is closer to living webs of people, ideas, and hard infrastructure. A future where we use Artificial Intelligence in how our systems interact, how our traffic flows seamlessly in a zero-emissions landscape where everything is carbon-negative and we have managed to develop energy forms that are rather regenerative and work together with nature. In fact, our world and cities look more like scenes from the movie Avatar rather than the concrete jungles that many world cities are. We live in a layered, multi-functional environment where we have mastered circular economy and we no longer have to worry about waste, given the new technologies that we have developed. These technologies and mechanisms enable us to recycle waste in forms that allow new ways of taking care of our ecosystems, and enhance our biodiversity.

Our infrastructure is adjusted to the rhythms of the sun, so that we capture all of the Sun's energy so that we waste none of it. Our bridges, roads, and buildings have the latest in-built, smart technology that focuses on maximising energy efficiency: our AI systems enable detection of problems before they occur and we have taught the algorithms to detect damages and alert the other systems that control e.g. traffic flows. We have also found new innovations from nature-based solutions where hard infrastructure has become entwined with nature: our buildings and bridges mimic nature by integrating trees, root systems, and vines to strengthen infrastructure, while delivering ecosystem benefits to communities like increasing shade, oxygen, and aesthetics.

With the increased impacts of climate change, our infrastructure must be adjustable. Increased storm activity has forced us to re-think how we design coastal settlements. Ideas such as embedded cyclone and storm shelters functioning as mini greenhouses where communities can remain safe from the storms while also having healthy ecosystems available for food production and drinking water. Other times shelter spaces act as gathering points for celebrations,

[†]Griffith University | j.nalau@griffith.edu.au

safe heavens in our cities where communities can relax and learn about their environment. Increased temperatures mean we design our spaces so there is enough airflow for an effective cooling system. In some cities, we have developed massive mobile cooling zero-emission fans that float in the air and adjust their position given the highest temperature locations. This way we are able cool our infrastructure in locations where the heat has increased beyond the ordinary.

We have developed a global Monitoring network that detects environmental degradation in real time, quickly showing us sources of any emissions that might be produced and keeps constant watch on our infrastructure functionality as well. Much like the brain, this centre is able to dispatch experts within communities and countries to help instantly if any ecosystems or infrastructures are starting to fail. Even if our cities have become automated in their functions, we have still managed to maintain human connections in these operations. We still live, laugh, and share common daily lives, histories, and cultures. We have become adjustable.

Biography: Dr. Johanna Nalau is an Adaptation Scientist and Australian Research Council DECRA Fellow researching the ins and outs of climate change adaptation at Griffith University, Australia. Her work examines and challenges the commonly held assumptions about climate adaptation (adaptation heuristics), which are rules of thumb that guide decisions on what climate adaptation is and how it should be tackled. She is a huge leadership fanatic and believes that good leadership enables better lives and better decisions across organisations and scientific fields. A robust future will be based on daring leadership that is open to innovation and has embraced future-back thinking.

— Infrastructure 2100 on the march! —

Edward J. Oughton and Ian Whybrow[*]

This beating heart of civilization,
Working effectively, sustainably -
telecoms and transportation,
plus water, waste and energy
and offers hope to every nation!

Thinking ahead can stop us feeling trapped
- efficiently, effectively -
when infrastructure can evolve, adapt.
There's no predicting every possibility,
But functionality and flexibility,
allows a decent future to be mapped!

Seamlessly one with a healthy society,
- earth-neutrally, harmoniously -
away with pointless infra-anxiety!
Let "plan humanely" be the norm
for moving people, goods and data
Let's reform!

So make a rowdy fuss! Let's have a proper scene!
Three cheers for Infrastructure in the Anthropocene!

---

[*]University of Oxford | edward.oughton@ouce.ox.ac.uk

Biography: *Edward J. Oughton is an infrastructure researcher with a particular focus on how to connect more people to a faster internet. Such information is vital for ensuring sustainable economic development as most new technologies require internet connectivity. His research is highly multi-disciplinary, drawing on analytical techniques from engineering and computer science, to answer questions pertaining to policy, innovation, planning, economics and sustainable development.*

Ian Whybrow a prolific writer having written over 110 books, translated into 27 languages, across 28 countries. His books are known for their humour and range from picture books to children's novels, short stories and poetry.

— Brief history of 21st century —

Francisco Pereira[+]

January 15, 2101

This article celebrates the 100 years of Wikipedia, and was automatically generated by automated historian, based on 2,567 internet articles.

0.1 2001-2031 - The great civilization decline

The early part of the first decade of the 21st century saw the long-time predicted
breakthrough of economic giant China, which had double-digit growth during
nearly the whole decade. In contrast, a global recession started with a housing
and credit crisis in the United States in late 2007, which led to the bankruptcy
of major banks and other financial institutions. The growth of the Internet
allowed faster communication among people around the world, and pushed
cultural globalization to establish itself. Social networking became a new way
for people to communicate no matter where they are.

Climate change and global warming became household words in the 2000s,
and during the first two decades, the world saw new record temperatures in different years and extreme weather events. The CO2 concentration rose from 390
to 410 PPM in the first years of the second decade, leading to the Paris Agreement
(2015). A youth movement was formed, inspired by the then charismatic teenager
Greta Thunberg. Predictions tools made significant progress and supported
studies, that influenced public support for political and economic investments
in countering climate change.

Between 2010-2020, information technology progressed further, and ubiquitous computing, particularly smartphones and tablets, became regular personal and household devices, later even more pushed by 5G broadband, which

[+]Technical University of Denmark | camara@dtu.dk

finally brought computing power to virtually every daily utility device (also called the Internet of Things, IoT). A drawback of all this technology was its quick obsolescence. New hardware or software versions were constantly being produces, eventually resulting in excessive electronic trash, waste of energy and resources.

0.2 2020-2031 - COVID19 and economic recovery

In the winter of 2019, the world faced its first large scale pandemic since the Spanish Flu of 1918, known as the COVID19 virus, which infected 55 million and killed 1.5 million people in 3 years. The unprecedented response in many countries, of ordering a lockdown for several months led to sudden shock to the economy, leading to an instantaneous crisis that took 7 years to fully recover. But more than the economic crisis, the pandemic revealed an extremely unequal world, created and constrained by the capitalist principles and aggravated by a wave of populist and corrupt leaders in several of the most populated countries. The globalization enterprise of the beginning of the century turned into a blame game, major alliances like NATO and UN faded, and countries siloed back into protectionist sentiment.

On the other hand, the world kept advancing technologically in many directions. Machine Learning model networks became everywhere, supported by 5G and federated learning algorithms. Models were no longer individual boxes, but instead networks of small models that could transfer knowledge, solve multiple tasks, and learn though their life cycle. This period was prolific in the embryonic large-scale modeling advancements that were essential for the True Democracy project.

0.3 2031-2048 - The True Democracy

During only 5 years, the world went through a sequence of massive near-extinction events, that culminated in drastic changes in society, particularly the way political decisions were made. It had become clear by the end of the 2020's, that political systems from Democracy to Dictatorship, and all the spectrum in between, would lead to one outcome in less than 20 years: mass extinction due to an accelerated breakdown of the planet's self-healing systems, the environment was collapsing faster than expected.

Even more than in the 20th century, the distance between decisions and real impacts (and between decision makers and affected populations) grew so large, that even the most modest decisions ended up being dragged by polarized political discussions - denier movements, toxic media influencers, proliferation of deep fakes and "relative truth" movements, always managed to destroy the discussion. Frustration of the populations grew dramatically with each natural disaster, while at the same time major technological companies in the world became more ingrained in all society levels - in fact, **they** were the ones in between decision makers and populations.

Strongly supported by the Esperanto corporation, the largest tech giant in the world, the near-defunct United Nations engaged in several successful small pockets of experiments in several islands in the world - the ones most immediately affected by fast sea-level rises. This was called "Project True Democracy". Only a few islands could save themselves, but a new hope was born.

In 2035, the "Miami agreement", the most important world agreement since the end of WWII, was reluctantly signed by the leaders of all nations in the town-hall of the last-standing floating citadel in the iconic city of Florida after a succession of 23 hurricanes in 2033. The Miami agreement consists of only 3 simple principles:

1. Sustainable environment future is an unalienable human right.

2. Both individual and societal freedom are subsumed to the appropriate balance of our planet, for the present and foreseeable future.

3. No human made changes to our planet are acceptable, that compromise the present and future of living species

4. Blatant violation of any of the above principles equates to aggression towards our species and our planet, and will be processed according to international court law

In the backdrop was always the "Project True Democracy", the blueprint for the practical implementation of these principles. More specifically, Esperanto's software, PLATO, which provided a "foreseeable future", allowed for "human made changes" that optimizes an "appropriate balance". More importantly, it provided the quantitative framework for principle Nr. 4, fully tested in Project True Democracy. PLATO was originally a government-funded project based

135

on a combination of research and technology developed since late 20th century, including: simulation, online gaming industry, behavioral econometrics, operations research, environmental engineering, and artificial intelligence.

In a PLATO supported society, citizens contribute their individual data, collected with their personal devices, which are implanted to prevent fraud. Such devices provide a wide range of individualized data, from biometrics to travel to social behavior. Data upload is done using federated anonymization techniques, which means that PLATO only receives parameter distributions from each citizen, never the data. These parameters summarize behavioral preferences at various scales (from households to neighbourhoods to country). Integrating into the general simulation, PLATO's AI explores the preferred spaces with millions of contextual variables (e.g. environmental, economical, social, geological) to propose decision making options.

Ideologically, the decision making options from PLATO can be ordered from left to right. Towards the left, we have those favouring the system optimum, while neglecting differences between citizens; towards the right, individualized customisation is considered, supportive of user optimum in exchange for sub-optimality at the system level. From the Miami agreement, it is implied that PLATO and its descendants constrain all components that disfavour environmental impacts - in other words, no decision making options are possible that affect human survival.

Citizen participation in PLATO happens in two simple ways. Every citizen has the right to recommend local or global interventions or policies, and every four years, each citizen has the right to vote on two options ("left" (0) or "right" (1)). Each option has associated an "executive" cabinet, comprising of people that are responsible for implementing all measures. The average of all voters will determine the decision making choice for the coming 4 years - a weight in a cost function - and the names of the chosen executives.

0.3.1 The IDI and the version 3.2 scandal

As per each individual's consent, the individual device implants (IDI - from LifeSense inc. - reads "eedee") could be used for well-being support, in terms of psychological stimuli (e.g. hormonal control, embedded mindfulness support) and nudging for more healthy habits in terms of diet, appetite, and physical exercise. Introduced to the mass market in 2033 (version 2.0), for more than 70% of the population, this functionality would be used at some point in life, often activated in teenage years and many years thereafter. Numerous studies and years of ethical evaluations proved the device's advantages, while enhancing the security provisions, introduced in version 3.0 (year 2040) and already tested with LifeSense under Esperanto in 2038 (LS had been acquired for a record breaking 2 trillion Euro). This new version brought about <u>DREAM</u>, a hypnotic method that allowed for continuous therapy while asleep, advancing a wide range of mind-enhancement tools. IDI3.0 with DREAM was used by psychotherapists and by schools of all levels, to provide learning-while-sleeping. The most popular functionality though, was enhancement of sexual experiences - users could choose hyper-realistic erotic dreams alone or with partners. DREAM quickly became pre-installed by default, and used by more than 95% of the world population above 12 years old.

The benefits of DREAM were recognized in 2047, when the Nobel peace prize was awarded to its team of creators for its undeniable role in absence of any conventional war in the world since 2040. Besides the benefits in tension release and healthy behavior, it was proven in randomized controlled trials in PLATO deployment contexts, that DREAM significantly increased general social interaction and communication skills.

IDI and DREAM's public perception suffered a dramatic hit in 2048, in what was known as the "version 3.2 scandal". In January 2048, IDI 3.2 was launched with minor updates, particularly in terms of security and quality of experience (e.g. sense of smell added to DREAM, <u>TIP</u> (Telepathy-over-IP)). This was during the 2048 elections campaign, in Germany, Russia, Brazil and South Africa. From DREAM's upgrade launch date to election day, October 30th, an estimated amount of 256 million IDI 3.2 users worldwide reported severe headaches. Only two days after the elections (all won by large majorities in either right or left), a free upgrade to version 3.3 was distributed and no more complains were reported.

In December, <u>UNA</u> (United Naturalist Association, an <u>NGO</u> comprising of IDI1.8 or below users), together with <u>PLATO-AI</u> in Esperanto, revealed a major backdoor security flaw in DREAM, that allowed for retuning personal long term memory - one could essentially redesign the "preferences" of people, in order to create new observed behavior. This had been exploited by a small corrupt group in Esperanto-IDI, sponsored by the major construction companies. While fraud on the data itself would be detectable, behavior change would be untraceable. When uploaded to PLATO, such preferences would lead to new large public investments. The problem was that this "retuning" process in some people was creating neurological and physical conflicts (e.g. related with personal trauma).

The backlash of massive downgrading of IDI versions and the appearance of a myriad of competitive devices eventually let to the restructuring of Esperanto, and the whole IDI product line philosophy (abiogenetic architecture, much less upgrades, much longer life cycles). Version 4.0 was launched in 2053, version 4.2 in 2076. Today, we use version 4.242 (a.k.a. IDI42), launched in 2082. Next year, IDI42 celebrates 20 years of existence, as the longest lived digital technology in History (as full-supported digital tool).

0.4   The gold rush of AI and the third AI winter

From 2017 to 2030, a period called the gold rush of AI marked a massive investment in Artificial Intelligence techniques, particularly fueled by large early 21th century corporations, like Google, Facebook, Alibaba or Huawei, and overwhelming public funding, particularly in China, United States and Europe. This started in the first two decades with advancements in Deep Learning, that were materialized into drastic improvements in areas such as computer vision, natural language processing, recommender systems, or autonomous mobility. Stimulated by the hype and collective imagination, many philosophers, opinion makers, and computer scientists recovered a discussion from the early 1960's AI days, on whether (or rather, when), would computer algorithms outperform our human intelligence. The concepts of Artificial General Intelligence and Singularity essentially postulated scenarios that would replicate the Turing test's fully capable human intelligence, in some cases even surpassing human intelligence. In extreme scenarios, this process would be very fast and dramatic, ending with human existence itself.

For corporations and governments, the promise of AGI power (and beyond) represented both a significant opportunity - for those being first - and a threat - for those second. This pushed even more investments in the direction of AI. However, as later proved in the 2028 seminal paper from Doe, Cabaza and Fuchs (2028), the Von Neumann's computer deterministic architectures of the 1960's, which separated processor, memory and interface, could not solve some of the key human intelligence tasks, namely figurative, associative, reasoning (e.g., metaphor, analogy, counterfactuals), h-creativity (historical creativity), even with the most sophisticated replicas of senses, memory decay, or limited cognitive awareness. While many of these tasks were individually replicated in the laboratory, or in sophisticated software tools for well defined interfaces (e.g. chatter bots, digital assistants, automated composers, painters), never could researchers build a single full-fledged multi-task machine capable of integrating all components together, with the emergent properties of consciousness, divergent thinking, serendipity or humour.

The failure of AGI preceded the third AI winter, but strengthened the role of two research communities (and respective funding resources) for the next two decades: quantum computing; organic computing. The former was based on quantum-mechanical phenomena such as superposition and entanglement to perform computation. Building on probabilistic programming developments (coming from machine learning developments between 2020 and 2030), physical implementations of analog quantum computers became an essential component of IDI and PLATO. Organic computing combined earlier developments from DNA computing (1994-2030) with material science developments, allowing for programmable material self-evolution, into general or specific purpose machines. As with Von Neumann architecture, both quantum and organic computing are full-fledged alternative computing architectures for general problem solving.

0.5   2048-2080 - Abiogenesis based population synthesis

The third AI winter pushed a massive crowd of prominent AI researchers and their teams to work on quantum and organic computing, which led to the emergence of the abiogenetic computing field. Based on early XXth century's in-silico architecture and the IDI version 3.2 scandal, the scientific community concluded the devices were too physical, cognitive, and sensorily unnatural for human consumption. The future was clearly abiogenetic, as DNA from the user was proven to be the best way to create symbiotic devices, and the only way to create an intelligent computer. As D. Libys said in her famous speach, "We only need to follow the instructions that lie within us from the beginning". Joan, the first intelligent abiogenetic computer, was publicly announced in 2051. Developed from Dr. Libys DNA, Joan was the fastest-growing organic polymeric cell replication technology of the day (from implant to human being replica in 3 years - a process that today takes 9 months). When she turned 5 years old, Joan was the first Turing-tested abiodroid. By the time she died (2058), there were 368,000 different abiodroids registered in labs around the world, including 151,500 from 37 non-human intelligent species (mostly primates, dolphins, octopuses and domestic animals).

In 2060, the genesis global coalition project, sponsored by the United Nations' Miami agreement high committee, unveiled the DNA re-coupling technique (DNAr), developed during the 12 years of the project's existence[*]. The technique allows for a biologically generated human to have a full resource-neutral life-cycle, with a simple DNA upgrade procedure within the first 4 weeks of pregnancy. The upgraded DNA changes the homeostatic control of energy of biological cells, allowing for a 100% diet on e-food (direct energy food[†]). Furthermore, together with IDI42, DNAr allows for 100% re-generation of a human being (or non-human being) to a given back-up point, in case of failure (e.g. accident, disease).

Thanks to DNAr's ethical evaluation from 2055 to 2063, it is impossible today to replicate DNAr (e.g. cloning) without losses. Each replication implies sharing of energy between clone and original, together with IDI42 memory conflicts. The naturalistic movements in the 60's and 70's against IDI42 and DNAr were gradually out worn by the benefits of these technologies to parents - the horror stories of naturalistic parents losing their children to the 2073 American flu (2073-nCOV) marked the end of an era.

0.6   2080-2100 - 20 years of peace

The last reported water uprisings in the Middle East and USA happened in 2074. Having many affected areas become inhospitable for non-DNAr citizens, the poles exodus of 2082-2096 led to the major known demographic changes in human History, although according to many sociologists and historians a major antecedent for the longest period of registered global peace, together with IDI42 rage/angst suppression and population growth control functionalities.

In 2099, UN's World Equality Organization (WEO) registered the highest rates of well-being and quality of life in known History.

---

[*]Corona et al, "DNA re-coupling through third level abiogenesis", Nature, 2060, check list of 987 co-auhors here

[†]Family of crystal composites, designed to look and feel like ordinary food, that can be generated by domestic e-blenders, that combine re-used crystals with energy, in a 99% circular economy cycle

Biography: *Francisco Pereira is Full Professor at the Technical University of Denmark (DTU) since August 2015, where he leads the Machine Learning for Mobility (MLM) group (http://mlsm.man.dtu.dk). He holds a Masters (2000) and PhD (2005) in Computer Science and Artificial Intelligence, from University of Coimbra (UC), Portugal. His methodological research combines Machine Learning and Transportation Research, and his preferred applications generally relate to transportation research problems, such as real-time traffic prediction, behavior modeling, advanced data collection technologies and transport modelling. He has contributed to top journals and conferences in both Machine Learning (e.g. IEEE Transactions on Pattern Analysis and Machine Intelligence, or AAAI) and in Transportation (e.g. Transport Research Part C, ISTTT), and thus lives constantly with his feet in both worlds, which he believes gives him constantly a different perspective, despite the hard challenges.*

He is also fed-up with the high-pace and short-sightedness of research competitiveness, especially w.r.t. impact factors, h-indexes, different colors of (paid) "open access" and respective embargos, biases on gender, color, age, PhD alma mater, and many other prejudices that distract us from actually doing proper science with fair resources.

— Urban infrastructure for a changing climate – dateline 2100. —

In 2100, cities' infrastructure in the post carbon future have adapted to address locally available urban ecosystem services and energy sources. Fossil fuel resources are very scarce and reserved for the highest and best social uses such as air evacuations, helicopter transported organs for urgent transplants, or disasters. The exact shape and configuration of the infrastructure will reflect local conditions – though all infrastructures will increasingly aim to buffer residents from growing numbers of high heat days, potential water, and food scarcities. The viability of long supply chains and global goods movement have all become more difficult, rendering the earlier infrastructure developed to accommodate enormous volumes of goods with huge warehouses and roads, anachronistic. Competing claims for urban space and its use (urban agriculture vs. solar generation for example) remain difficult and constantly negotiated.

But, overall, the lack of fossil fuel energy has required cities to become increasingly self-reliant and connected to nearby hinterlands for resources. Within cities, infrastructures have important roles to shield residents from the extremes of climate change, from heat to inundating rainfall and floods, and sea level rise. These infrastructures are now gray/green, working with nature to enhance its local benefits. One of the major assets of cities – streets and sidewalks, building setbacks – has dramatically transformed. They have been reconfigured to infiltrate stormwater. Saving rainfall has meant that streets, sidewalks and open spaces are sponges and storage facilities where there is no groundwater capacity. These are widely deployed, above and below ground. All roof tops or canopy covered areas have water capture that drains into groundwater, barrels or cisterns. Hard impermeable surfaces are a thing of the past, and the biggest challenge in high rainfall periods, is ensuring surfaces don't become impassable muddy pathways. Urban landscapes have been converted to reflect the climate regime. In hot dry climates, decorative plants that can withstand long

---

†University of California, Los Angeles | spincetl@ioes.ucla.edu

water-less periods are the predominant species, saving outdoor water use for the community gardens, where that water is carefully managed. In cooler climates, plants are used to buffer winds and weather, and to cool/shade for high heat periods.

Streets, and buildings capture sunlight for solar power. With solar canopies over streets turned into walkable, bikable networked pathways, the converted streets offer shaded transit corridors where solar generation is possible for both electricity and hot water. Other less appropriate streets, due to orientation, benefit from shade structures and are punctuated by leafy appropriate trees at well thought out locations that include public amenities, whether fountains, play areas, chess or ping-pong tables or other. Cities are warrens of shaded alleyways and streets, keeping the atmosphere cool. Few private vehicles are in use and none use fossil fuel energy. Wider streets are covered by solar canopies connected to microgrids.

More dramatically, the era of the single-family zone has ended. In a constrained 21st century era, the single-family zone stands out as highly vulnerable to increased heat, water intensive, isolated and inefficient. Its car-dependency is no longer sustainable due to the reduction in fossil fuel consumption and the competing needs for renewable energy such as for electricity for building, making for abandoned or greatly densified former suburbs. Pen spaces patches – like back yards – however have been preserved and repurposed for food production, including intensive fish farming. The front setback may accommodate small local service businesses, a small grocery store selling or distributing locally grown produce, fish, eggs and poultry or rabbits, or offering other services. Large animal protein is rare and dependent on the capacity of the local hinterlands to grow large animals with no additional feed that must be processed and transported.

The scale of infrastructure and services has shrunk to reflect human proportions and energy limitations. For example, multi-story buildings have few elevators, so their height is limited. Intense energy resources – and some fossil fuels– are reserved for high priority places such as hospitals, police and fire stations, but the rest of the city has vastly reduced its energy consumption to scale with available energy. A clever array of different electric, human/electric vehicles are used for goods transport and any needed commuting, as human energy often needs to be deployed. Neighborhoods are compact, and space – like current building setbacks – is used intensively for food, fiber, locally needed services. There is no wasted space, only purposeful space, that can include leisure and beauty, but there is no space that is simply overlooked or not considered.

Appliances, whose manufacturing was extremely energy intensive, are now small, exceedingly well made with repairable components. They are also highly energy efficient. The grid has been retooled to provide DC power for many appliances, saving energy and suited to locally generated power. AC power is reserved for essential uses. For regions with geothermal, the general 20th century district heating infrastructure has been retooled to make up for the lack of additional fossil fuel energy, and in colder climates with forests, wood energy is used in supplement to solar. Wind energy resources are utilized as well, but with climate uncertainty and violent weather, dependency is strongly assessed against vulnerability to disruption, damage and disaster.

Repurposing, reuse, recuperation, tight coupling of manufacturing with locally available resources and reduced scales of infrastructure emerge, offering locally creative and specific solutions to place. Infrastructure reflects available materials, climate, and human/ecological partnerships in which the use of nature maintains its health and long-term viability. This is a zero-waste society of deep partnership with and understanding of locality. Gone is the one-size fits all approach to infrastructure, once made possible by virtue of high energy-dense fossil fuels and the seeming limitless Earth resource base. Rather, solutions are autochthonous, responses to specific problems in specific places, and aimed to address the common interest.

Biography: Stephanie Pincetl is Professor and Director of the California Center for Sustainable Communities at UCLA. Dr. Pincetl conducts research on environmental policies and governance and analyzes how institutional rules construct how natural resources and energy are used to support human activities. She is expert in bringing together interdisciplinary teams of researchers across the biophysical and engineering sciences with the social sciences to address problems of complex urban systems and environmental management.

— A wake-up call —

Maria Pregnolato[†]

2100, November. Monday is always Monday.

I wake up: a green light reflects on my wall mirror, my alarm clock alerts me to a red warning for air quality. Opening the windows is not recommended. I go to the toilet, open the shower and let the water run. I have a hearty breakfast with bacon, milk and eggs and I am ready for the day.

Fig. 1. A parking space, aerial view (photo by R. Searle on Unsplash).

My workplace is 60-km away and my SUV will get me there. I am late as usual, and traffic is a problem of which I am not part of. Building a new lane has not

[†]University of Bristol | maria.pregnolato@bristol.ac.uk

improved the flows, but I really appreciate the robots that bring you coffee. It's 40°C and I am freezing in my car thanks to the air conditioning. I sip the coffee and throw the plastic cup out of the window, just on the track of the old railway. This popular habit has built coaches of trash on the tracks.

I arrive at work and I park my car in bay no. 2354, in the C parking building. 50,000 m²of concrete is facilitating the connections between parking, offices and services. It's nice and flat, ideal for the automatic moving walkway. My personal assistant activates and reminds me about the meetings with the Russian partners and the Antarctic delegation. My company is trying to build a consortium for designing a new industrial complex in the Amazon.

2100, November. Monday is always Monday.

I wake up: a pale sun reflects on my wall mirror, my alarm clock reminds me to walk 10000 steps. I open the window and appreciate the fresh air. I go to the toilet, get in the shower and sing a song. I have a hearty breakfast with soya yoghurt and rhubarb (from my allotment), and I am ready for the day.

Fig. 2. Solar panels for producing green energy (photo by A. Gücklhorn on Unsplash).

My workplace is 10-km away and my bamboo bike will get me there. I am late as usual, but I still prefer my bike to the new solar light rail. The bike path has replaced the old highway and it is very safe - including at nights, thanks to the new self-illuminating pavement. It's 10°C and pedalling keeps me warm. I sip water from my backpack, which I will refill at my office at the rainwater fountain. The habit of recycling, together with investment in recycling infrastructure, has reduced waste to a minimum.

I arrive at work and I park my bike near the acre of wood, opposite the beehives. I talk with my colleagues while I climb to the 15th floor. The supersonic lifts are used only by those who need to. I just have enough time to do my 30 daily jumps on the energy-absorbing platform. I opt to walk during the meeting and dial in to the boardroom. My company is planning to 3D-print a biomimetic hospital in the previously industrial area of the city.

2100, November. Monday is always Monday.

Which future would you like to wake up in?

Biography: Dr Maria Pregnolato is a Lecturer in Civil Engineering and EPSRC (Engineering and Physical Sciences Research Council) Research Fellow at the University of Bristol (UK). She has been a visiting Assistant Professor at the University of Washington, Seattle (USA) in the 2019 summer. She investigates the impact of flooding on the built environment, especially the risk and resilience of urban infrastructure. She studied in Italy and China within the Italian-Chinese curriculum of the University of Pavia and the Tongji University of Shanghai during her MSc in Civil Engineering-Architecture; she began her academic career at Newcastle University (UK), where she developed an integrated flood-transport model to explore the impact of flooding on road networks. Maria believes that science has a crucial role in developing solutions to global change and demand, by improving the resilience and the quality of our cities through applied research.

Acknowledgements: the author would like to gratefully thank Liz Lewis and Fulvio Lopane for edits and suggestions.

— Puzzling Over Our Mobility Futures —

Megan S. Ryerson & Carrie S. Long[†]

Puzzling Over Our Mobility Futures

When governments and agencies collaborate and plan for future infrastructure needs and use as one big system, we will serve, restore, and revolutionize mobility for 2100, thus avoiding a status quo of increasing mobility fragmentation.

2020 Mobility as a **Puzzle**

- Pieces are planned individually
- Pieces compete to fill the same needs while others are left behind
- No one piece wants to lead or take responsibility for integrating the network

Beyond 2020

Status quo is preserved

Agencies and institutions collaborate and work to integrate their systems.

Infrastructure is planned thinking of how travelers and shippers do and will use the system.

Mobility as an **OBSTACLE**

- Transportation systems are built to optimize the narrow goal of each agency and operator
- **Equity Concerns:** Travel over the fragmented network requires deep navigation knowledge and physical ability
- **Environmental Concerns:** As individuals continue to optimize their own travel on systems that are not priced to reflect externalities, environmental emissions grow
- **Quality of Life and Congestion Concerns:** Infrastructure demand and supply is deeply uneven; extreme concentrated congestion and underutilized areas

Mobility as an **INHIBITOR**

- **Equity Concerns:** The gap between those physically and economically able to traverse the fragmented system widens
- **Environmental Concerns:** Environmental externalities cannot be remedied and infrastructure is extremely vulnerable to climate-related inundation
- **Quality of Life and Congestion Concerns** Transportation is an impediment to the quality of life for large groups of citizens

Mobility as a **SYSTEM**

- Shared, integrated frameworks
- Travel is planned and optimized as a comprehensive whole
- Infrastructure priority and funding is based on service and access

Mobility as a **SERVICE**

- Transportation is designed to serve customers' needs
- Travel is personalized — trips are point-to-point journeys, with multiple modes; journeys are as frictionless as possible
- Travel is integrated with Mobility-as-a-Service payment structures

Mobility as a **RESTORER**

- Travel is more considered, and costs (i.e. fares, tolls, and taxes) reflect the true cost of travel and its environmental impact
- Obsolete infrastructure is reclaimed for new, sustainable and equitable uses such as green and shared public space

Mobility as a **REVOLUTION**

- Infrastructure changes and advancements impact how and where we choose to live and how we use the public realm
- Shared travel and shared modes are the norm, not the exception

Carrie S. Long, AICP
Megan S. Ryerson, Ph.D.

[†]University of Pennsylvania | mryerson@upenn.edu

Biography: Dr. Megan S. Ryerson is the UPS Chair of Transportation and an Associate Professor of City and Regional Planning and Electrical and Systems Engineering at the University of Pennsylvania. She received her Ph.D. in Civil and Environmental Engineering from the University of California, Berkeley in 2010 and her B.Sc. in Systems Engineering from the University of Pennsylvania in 2003. Dr. Megan S. Ryerson founded the Center for Safe Mobility in 2018 with the goal of developing new methods and metrics for human-driven transportation safety studies.

Carrie S. Long, AICP is a Transportation Planner with ambitions to rethink the design and use of streets. As Director of the Center for Safety Mobility, Carrie manages multidisciplinary research projects focused around transportation infrastructure design and safety. Carrie received her Master's in City Planning from the University of Pennsylvania and her Bachelors of Science in Industrial-Organizational Psychology and Bachelors of Arts in Sociology from the State University of New York at New Paltz.

— Detours and Funiculars: Towards sustainable urban transport infrastructure in 2100 —

Shoshanna Saxe[+]

To start the game:
What is the dominant mode share where you live?

(a) Automobile: Start on square 1.

(b) Active or public transport: Start on square 20. You have a head start towards sustainable transport in 2100!

Pick a small object to be your representative on the board. Place object on your starting square.

On your turn:

Roll one 6-sided die. Move your object the number of squares you roll.

If you land on a funicular entrance, congratulations you are building and investing in the infrastructure of the future - ride the funicular towards sustainable transport infrastructure in 2100.

If you land on a detour entrance, oh no you've made an infrastructure choice that takes you away from the needs of 2100 - you must take the detour backwards and hope future choices lead you in the right direction.

Winning the game:
Take as many funiculars and avoid as many detours as possible. Make it to square 100 (the year 2100) with sustainable urban transport infrastructure, and win! The more players who make it to square 100 the better, every win multiplies the impact of the other wins. After you reach square 100, encourage your

[+]University of Toronto | s.saxe@utoronto.ca

fellow players to aim for the future by using your extra turns to provide rolls for the others (roll the die and offer the number to another player). Just like advice, the players can ignore your offered roll if they like, especially if you are trying to land them on a detour.

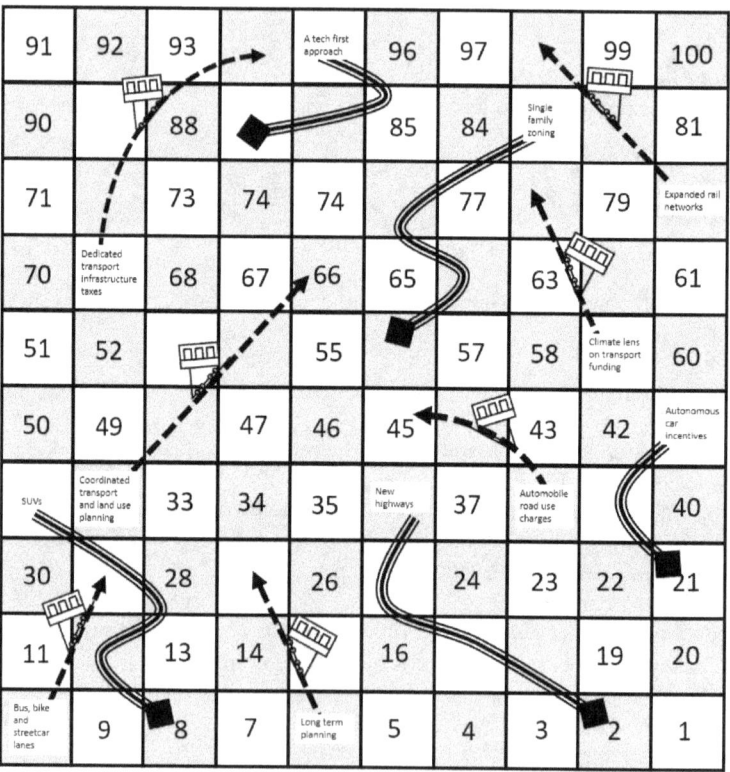

91	92	93		A tech first approach	96	97		99	100
90		88			85	84	Single family zoning		81
71		73	74	74		77		79	Expanded rail networks
70	Dedicated transport infrastructure taxes	68	67	66	65		63		61
51	52			55		57	58	Climate lens on transport funding	60
50	49		47	46	45		43	42	Autonomous car incentives
	Coordinated transport and land use planning	33	34	35	New highways	37	Automobile road use charges		40
30	SUVs	28		26		24	23	22	21
11		13	14		16			19	20
Bus, bike and streetcar lanes	9	8	7	Long term planning	5	4	3	2	1

Biography: Dr. Shoshanna Saxe is an Assistant Professor in the University of Toronto's Department of Civil and Mineral Engineering. She investigates the relationship between the infrastructure we build and the society we create to identify opportunities – and pathways – to better align infrastructure provision with sustainability. She has been recognized by Clean50 as one of Canada's emerging environmental leaders and was awarded the 2019 Ontario Engineering Medal – Young Engineer. She specializes in thinking holistically about large civil infrastructure systems.

— Mobility and the City in 2100 —

Yoram Shiftan[†] and Alona Nitzan-Shiftan

Mobility technology is accelerating ever more rapidly, far outpacing what arrangements cities put in place to accommodate them. From where we stand now, we can quite safely forecast what kind of exciting technologies will be available in 2100. But we still know very little about how they would integrate into existing infrastructures and into new cities. What happens when these technologies spill over from research universities and high-tech centers into our cities? How can we harness them to improve our cities, to interact with existing urban forms, institutions, and markets to the benefit of the city's residents?

To envision the city in 2100, we need to first qualify the question. We need to understand that we should not ask only how technology can change our cities, but, more importantly, we need to ask what is the city that we want to create by implementing cutting edge mobility services? If we agree that technology is the means and not the goal, then innovation should result from comprehensive planning that goes beyond the challenge of writing algorithms to activate sensors or coordinate data. More than the city of unending technological thrills, we need a city of an upgraded everyday life, one that makes sustainable and equitable use of resources, and a society where investment in the public realm is monitored through democratic checks and balances.

The future city we envisage in this chapter would be configured by urban designers, urban planners, and policy makers. They would adjust the physical form of the present-day city and specifically the urban infrastructure we know today, to accommodate new mobility technology in ways that ensure the best livability of its people. We start with a brief history of future cities that reason our restrained utopia. We then outline the available mobility technologies in 2100 and follow up with the focus of this chapter—an illustration of an urban center of a metropolitan area and its transport system. We conclude with the

---

[†]Technion, Israel Institute of Technology | shiftan@technion.ac.il

policies and behavior of travelers that keep the city vibrant, sustainable, and just.

The Future City in History

The promise of automated systems that would move people safely and rapidly to their desired destination, on ground and in the air, with as little effort as pressing a button, ignites the imagination with futuristic images. Such were the images that were already produced in the early twentieth century when the fruits of the industrial revolution had begun seeping more and more into people's lives. Avantgarde artists such as the Italian Futurists adored the machine, imagined sublime industrial compounds, and were ecstatic about the prospect of motorways. American visionaries such as Hugh Ferriss drew fanciful spatial infrastructures that wrapped around mega skyscrapers hundreds of stories tall. The legendary Frank Lloyd Wright restored libertarian values in the Broadacre City he envisioned in 1932, a car-centered utopian anti-city that foreshadowed the postwar urban sprawl toward an ever-growing suburbia. Most spectacular was the Futurama that Norman Bell Geddes designed for General Motors in 1939 for the New York World Fair. He circled the public around a huge model of futuristic urban landscape that demonstrated how one can drive through inter-city highways, from coast to coast, with no traffic lights.

In Europe most influential was the Radiant City of Le Corbusier, a French modern architect and urban reformer. He boasted a landscape of residential towers that he orderly distributed in the open space in order to devise a hygienic, well-ventilated and car accessible urban environment. He even argued that authorities wipe out *La Rive Droite* of Paris in favor of his orderly built vision. Le Corbusier was a leading player in the seminal International Congress of Modern Architecture (*Congrès International d'Architecture Moderne*, [CIAM], 1928-1959) that responded to the overcrowded city of the industrial revolution with strict zoning guidelines. The urbanism of the Modern Movement separated the city to four major zones—work, residence, leisure and circulation. The eventual global reach of this zoning vision assumed its most dramatic embodiment in Brasilia, the new capital city that Brazil erected from scratch in its hinterland. When it was populated during the 1960s, the urban ideology that underwrote Brasilia's plan was already falling out of vogue.

Indeed, when the means to execute these modern visions were available, their drawbacks were painfully experienced. The promise to improve urban life by means of ordered and zoned city prompted modernizers to bulldoze ostensibly dilapidated mixed-use urban habitat in favor of residential towers surrounded by inner city highways. The mid-century sweeping industrious urban renewal extravaganza met prominent critics like Jane Jacobs, whose *Death and Lives of American Cities* wittily insisted that the city belong to its dwellers. The advocates of the city refused to surrender to new technologies that threatened to throw the baby out with the bathwater.

Undoubtably we are now in the midst of a new revolution of no lesser magnitude. Digital and information technologies and new automation capacities are transforming our lives and our cities as much as the motorized car and electricity had in the twentieth century. The question is how to learn from the former 'industrial revolution,' and to improve the city without losing sight of the things that people have appreciated in cities throughout history. How do we update our urban landscapes while also keeping a mixed-use city, human-scale streets, safe and livable public space and casual human interaction? To meet this urban challenge, we should review the future technologies that would enable the city we envision.

The State of Technology in 2100
Let us move to this future city. The development of its new transportation environments is contingent on the rapid development of advanced technologies, such as Information Communication Technologies (ICT), Internet of Things (IOT), sensing, electrification, and battery technologies. Most pertinent are automation and connectivity technologies that enable efficient driverless transportation services, and thus may reduce or eliminate the need for private vehicles. It is important to keep in mind that automation and connectivity are two different technologies, however. After starting off by describing these two, we introduce emerging travel modes currently under development, and conclude this section with a discussion about how all these technologies create a new word of travel experience knows as Mobility as a Service (MAAS).

Automation
In 2100 all vehicles are already fully connected and automated (henceforth: CAV, connected automated vehicles.) The levels of their development, according to the Society of Automotive Engineers (SAE) range from 0 to 5. In 2020 the industry has already succeeded to produce and test fully driverless Level 4 cars, but these cars can operate only in designated areas with ideal road condi-

tions. By 2100 level 4 vehicles will dominate the market. At the current stage, the question whether level 5 cars, which can operate anywhere and under any conditions, will overtake them by 2100 is still unclear. To be sure, the future city that we discuss later in this chapter can be advanced with level 4 vehicles and does not depend on the full operation of level 5.

Connectivity

Connected vehicle technologies allow vehicles to communicate with each other (Vehicle to Vehicle, V2V) and with the world around them (Vehicle to Infrastructure, V2I). In 2020 technologies have already helped drivers making informed decisions and allow transportation agencies to better manage traffic flow. In 2100, all vehicles are fully connected and automated, and are operated in city centers under a central transport control center (TCC). The TCC navigates all moving elements within its orbit and can therefore ensure efficient flow without the use of traffic lights and signals. Special procedures have developed to ensure travelers' privacy by separating travelers' identity from the travel mode they use.

Urban Air Mobility (UAM) and More

An addition to city life is air robot taxis that are available for personal use. They can land and take off vertically in a relatively small area without a runway. They are electric and use multiple rotors to minimize the noise of rotational speed. Other fast travel modes include sky-trains that are carefully designed to connect skyscrapers without hovering over the city too heavily. At the pedestrian friendly end, we find automated personal pods, as well as people movers.

Mobility as a Service (MAAS)

Jointly, these technologies enable an efficient transportation system known as Mobility as a Service. MAAS is a technology platform that coordinates various transportation services and options, from both public and private transportation providers, under a unified gateway. Travelers can use the platform using a dedicated app to create, book, and manage trips, as well as to pay for trips using a single account. It provides access to all travel modes, ranging from rentable e-scooters to metro—including normal-sized cars, single seaters, double seaters, and aerial robot-taxis for private or shared rides—and manages the transfer between them. The app will offer travelers different options for a requested trip that vary by travel modes, travel time, comfort, and price.

The app navigates the user along a preselected route but adjusts it as necessary. It is a sort of a transportation marketplace. As a highly integrated and convenient system, the MAAS provides travelers with a variety of personally tailored travel options that improve travel experience to the point of making private car ownership redundant. The more connected CAV is, the more efficient and cheaper MAAS becomes.

The City Landscape in 2100

The city we envision is the inverse of CIAM's modern urbanism. It is a city that uses its technologies to enhance walkability between its intentionally diverse and mixed uses. The zoning of our future city does not divide it: quite the opposite. It focuses on a wide array of traffic modes and speeds and ensures connectivity between its different functions.

The twenty second century city prioritizes people. The viability of pedestrian life is the focal point from which all other considerations unfold. Accordingly, the city's main urban spaces, which are designated as Zone A, allow uninterrupted walkability. Only slow mode transportation, such as e-scooters, pods, and people movers, can share this space with pedestrians on designated micromobility lanes. Pedestrians can use sharable personal and dual automated pods, and special pods assist elderlies and people with disabilities (as in airport today.)

Zone B accommodates both pedestrians and limited speed vehicles. This includes micro-transit, dynamic ride hailing services, and various forms of fixed route public transport, including trams and light rail. Zone B is divided into two sub zones. In B1 pedestrians are prioritized, a special emphasis is placed on accessibility, and vehicular speed is limited at 20 MPH. In B2, mobility is on par with accessibility and vehicular speed limits are set to 40 MPH. Zone C is dedicated only to fast movement of both vehicles and mass transit. It consists primarily of an advanced underground metro system, but also of air mobility routes and sky-trains. Accordingly, its main operation is either under or above pedestrian ground level, and it is therefore fully separated from pedestrians.

Figure 1 depicts the future of an existing city, rather than a simulation of a new city built from scratch. The new technologies are interwoven into the existing building surfaces, street spaces, and transportation infrastructures of the city in a manner that respects the city form we know and cherish—major streets and

Fig. 1 - Perspectival cross section of a street in 2100 with different modes of urban mobility. See all figures in high resolution at https://doi.org/10.6084/m9.figshare.12735491.

boulevards that envelope buzzing commercial activities. The ground level of the main street prioritizes pedestrians and follows the regulation of Zone A. This is where one goes for shopping, frequents coffee places and restaurants, participates in social gatherings, rests in pocket gardens, and fetches an e-scooter or a pod to reach a desired destination. The livability of this main zone is contingent on the easy and unfettered access that new modes of transportation afford.

Fig. 2 – Vertical travel modes connecting between zones.

The affordability of access is crucial. In 2020, there are extended pedestrian malls in large city centers, but they are blocked from most traffic modes and are therefore very limited in size. In our future city, vehicles can move both horizontally and vertically and can therefore gain access to an extended three-dimensional space. This means that the traffic zones we describe are functionally separated but spatially connected in the same urban area. In other words, Zones A, B, and C can simultaneously provide access to the center through very different modes and speeds of transportation.

The success of our future city is contingent, to a large extent, on the seamless transition between the different zones and modes of transportation managed through Mobility as a Service (MAAS) technology. Examples detailing such transitions are shown in two smaller illustrations that detail sections of the larger picture.

Fig. 3 – The metro and its vertical connection to the Zones C, B, and A.

In Figure 2 we see the connection between Zone B1, which pedestrians share with limited speed vehicles, and Zone A, which is predominantly pedestrian. The public transit we see in Zone B1 is composed of several linked CAV vehicles. Upon request, one vehicle detaches from the chain, moves sideways and hooks to an elevator that raises it to Zone A. People alighting from the vehicle can take a pod or e-scouter that bring them to their destination or walk. In specific buildings, CAV can move vertically to higher levels, and potentially connect to sky-trains or air taxis. Once the car descends back to Zone B1, it can hook back to the multi car transit.

Figure 3 shows the upgraded metro system that runs in the pedestrian free and fully automated Zone C. The twenty second century metro is still the most efficient mass transit mode in the city, but it is remade to be exceedingly user friendly. Its automated trains are constructed as solid, heavy cars within which nestle lighter inner cabins where people sit. In major stations, these inner cabins detached from the train and ascend to Zones B and A, where people alight *en masse* from the train directly onto the city's main pedestrian center. Once new passengers board this elevated car, it descends back with the same elevator until it reaches the heavy train that awaits it in Zone C.

A renewed look at a B&W version of figure 1, illustration 4 shows how the moderation of speed in Zone A and B allows for a greater design flexibility of the city's public spaces. Multi-level spaces spatially weave several urban planes, below and above the main street. Likewise, illustration 4 shows us how the different traffic zones are not identical to urban levels. Instead, they are distributed on various levels, from Zone C metro, highways, and sky-trains, to level B of slow CAV modes, to the predominantly pedestrian Zone A in a way that encourages mixed use facility and rich urban experience. For those of us who remember the private car city, with its heavy traffic and long rows of parked vehicles or parking garages that occupy the cherished seam between the pedestrian area and the buildings around it, the expanded public space that is serviced by transportation rather than occupied by its needs, is a breath of fresh air.

Epilogue

New technologies of automation, connectivity, and MAAS promise to create multiple new services and improve travel experience. The question how exactly they will impact our cities and quality of life still remains, however. It is difficult to resist the temptation of diverse mobility in urban settings, but it comes with a price. On the bright side, it provides people who cannot drive or who face difficulties in accessing public transportation services (children, disabled persons, the elderly) with markedly increased mobility; travel becomes much less time and energy consuming, as people can freely engage in other activities while "driving"; and the absence of a driver and streamlined infrastructure reduce the price of travel services. The result is a better all-round travel experience, that may in turn encourage more and longer travel, which is expected to increase congestion. We know that the introduction of ride hailing services, like Uber and Lyft, has already increased congestion in large US cities in 2020. Such services attracted more public transportation passengers than private cars users. CAV and MAAS are expected to exacerbate this trend. How, then, should the local government and the residents of our future city address this challenge?

The government of our future city has significantly and continuously invested in mass transit modes that it estimates are as the most efficient for large dense cities. It allows alternative travel services to complement mass transit for the last mile solution and inner-city accessibility, but also realizes that they cannot hope to achieve a similar volume capacity as the underground metro services

Fig. 4 - The distribution of mobility zones in the city.

without disturbing urban life. People are free to choose their preferred travel mode from multiple options that best meet their needs and circumstances. At the same time, they accept the necessity of managing the demand for trips.

This is done by allocating the road space in the city center primarily to public transport services and developing an advanced real time road pricing system. When a single passenger occupies a car, there is a social cost of delaying all other passengers, thus road pricing is determined by congestion and occupancy. The larger the space a person occupies, the higher is the charge for the ride. Air taxis are particularly expensive—intentionally taxed highly in order to prevent mass use, and to clear the sky above the city of congested movement and visual disturbance.

The various charges allow the government to generously subsidize an affordable mass transit for all residents. As a result, people in our future city have also modified their cultural preferences and adapted their travel behavior. First, the highly reliable travel services convinced them to substitute private ownership of a vehicle with the MAAS system. Second, they understand and accept the price for using a road: the more space you take on the road, the more you pay. Once people have accepted and supported this policy, the opposition to road pricing declined. With plummeting numbers of individual car ownership, it becomes difficult to recall why would anybody object paying for such rudimentary services. Once MAAS provided a better and more affordable option for increased accessibility and flexible mobility, the age of the privately owned vehicle has come to an unexpectedly swift close.

Acknowledgments: All illustrations were prepared by Amit Sadik and Roy Schneid, graduate students of architecture at the Faculty of Architecture and Town Planning, the Technion. The authors thank Nir Sharav, Yuval Shiftan, Iris Aravot and Zvika Koren for their insightful comments.

Biography: Yoram Shiftan is a Professor of Civil and Environmental Engineering at the Technion specializing in travel behavior, transport policy, and transport project evaluation. He is the head of the Israeli Center for Smart Transportation Research and was the head of the Technion Department of Transportation and Geoinformation Engineering and the head of the Technion Transportation Research Institute. He was the editor of the journal Transport Policy and the chair of the International Association of Travel Behavior Research (IATBR). He received his Ph.D. from MIT and since then has published over hundred papers and co-edited five books. Overall, he spent 4.5 years as a visiting professor in leading universities in Europe and the US, most recently at Northwestern and University of Illinois, Chicago.

Alona Nitzan-Shiftan is an architectural historian and theorist, and an associate professor at the Technion, where she heads the Arenson Built Heritage Research Center. She was the first chair of the Architecture Department at the Technion and served as the president of the European Architectural History Network (EAHN). She received her PhD from MIT, and her work on the politics of architecture and heritage, on architectural modernism in Israel and the United States, and on critical historiography was sponsored by CASVA, Getty/UCLA, and the Frankel Institute at the University of Michigan. Her awards winning book Seizing Jerusalem: The Architectures of Unilateral Unification was published by University of Minnesota Press. She currently works on the Israeli volume of Modern Architectures in History (Reaktion Books).

— An Off-Site and Off-Grid Urban Future?
—

Joel A. Tarr[*]

In examining past urban infrastructure transitions to gain insight into future patterns, one notes how the literature is dominated by the rise of networked urban infrastructure. During the 19th and early 20th centuries American cities transitioned from walking cities, offering minimal services to their inhabitants, into networked entities that provided water supply systems and sewers, railed streetcar lines, electrical power, street lights, and telegraph and telephone communications. Thus, cities appear to have moved inexorably towards what Steven Graham and Simon Marvin label the "modern infrastructural ideal" of providing uniform infrastructures largely through technical networks.

A closer look at the historical pattern, however, presents a somewhat different picture. Movement towards networked systems was not necessarily uniform. Networks were not integrated and existed in silos, often causing negative interactions and feedback. Households and firms frequently had off-grid and on-site provision of services such as water supply, waste disposal, and individual gas and electrical producer units generating power for homes and businesses. The built environment of cities was marked by the co-existence of infrastructure networks and off-site devices providing services, often of a similar kind. Thus, the built environment of 19th and 20th century cities can be characterized as marked by mutations, or new devices; recombinations, where old and new technologies were applied from one area to another to create a new artifact or infrastructure; and hybrids, which combined different and new inventions in a novel way.

As we project the built environment of future cities, we can expect to find a mixed landscape of networked systems, on-site and off-grid facilities, and hybrids, all providing services. Because of the increasing vulnerability of networked energy systems (the grid) to disruption as well as their environmentally costly fossil fuel dependence, we can forecast an increasing reliance on off-grid devices such as solar power in the energy domain. Some areas, such

[*]Carnegie Mellon University | jt03@andrew.cmu.edu

as the water supply of large cities, will undoubtedly remain largely centralized and networked because of issues of public health and the economies of scale. On the other hand, in locations marked by limited rainfall and frequent drought, ground water and rainwater capture will be increasingly drawn upon, often from individual household wells with decentralized treatment technologies. In addition, decentralized and improved waste disposal systems will become more common, replacing piped sewage systems whose waste streams require extensive sewage treatment before disposal. In the area of storm water control, instead of building large grey infrastructure pipes and retention basins, cities will increasingly rely on green infrastructure to retain it on site. And, provisions for the safe and convenient use of city streets by pedestrians may supplant roadways now oriented towards the automobile.

The "Smart City" of the future, therefore, will in some ways not only be filled with "smart" innovations but will resemble a version of the city consisting of a mix of older and newer methods of service provision rather than a completely modernized landscape. As historian David Edgerton has maintained in *The Shock of the Old*, millions of people continue to use older technologies over new "innovations" because they meet their needs. Familiarity and comfort may be more important values to them than innovation and modernism.

Biography: Joel A. Tarr is the Richard S. Caliguiri University Professor of History & Policy at Carnegie Mellon University. He has authored numerous essays and books about urban infrastructure and urban environmental issues.

— The Untraceable City —

Frances Taylor & Paul Hoekman[†]

"Gran"

"Yes Bunkorbio?"

"Why does this Valentine card from grandpa have a thorny weed on the cover? Did you make him angry? Was it a joke?"

"No Sweetie. Back then people loved those 'weeds'; they are called roses. People would buy roses from all over the world on Valentine's day to show romance. He was telling me he loved me very much."

"Were you one of those queens or something that he imported something from overseas for you?"

"How old do you think I am Bunkorbio? Do I look like a queen? Queens were more common a whole *hundred* years before I was born. Queens were not very common at all when I was alive. The last few were kicked out of even symbolic power a number of years after I was born."

"So how could grandpa send you weed flowers on Valentine's day?"

Shivo: "They used planes for that Bunki. That's why we need to speed things up again, so that we can also get things quickly."

"No Shivo, I will not tolerate that kind of talk from you. You have no idea what it was like back then. You have no idea. Pfft! Sit down. Let me tell you what it was like, you little worms. You have to know how things were to know what a bad idea it is to start The Great Rush back up again. And to know how we got into the global habit of importing perishables. What silliness."

[†]Communitree (Frances) - International Society for Industrial Ecology, Metabolism of Cities (Paul) | frances@communitree.in

Shivo: "But why is importing just a few things, and speeding things up just a little bit such a bad idea? Why wait for so long to eat strawberries when I could have them on my birthday next week? Xhali is so lucky her birthday is in spring. She has strawberries every year. Why can't I have strawberries on my birthday? It sucks!"

And that is exactly why we don't let greedy impatience rule our decision making. Sit still for a few minutes and let me tell you what it was like.

"I was born in the year 2010 when the Great Miswanting and The Great Rush were starting to crack apart. Do you remember when we visited the Museum of Capitalism? We saw these paper sheets called money."

Bunkorbio: "I remember, it was used to choose which people were better than others."

"Well, sort of. It wasn't really that people were chosen, but instead it was based on where or when you were born, who your parents and family were, or luck. And sometimes on how hard you worked. But remember, money was used so that people could own things. Someone would own a house, or many houses. Or a car, a bicycle, and even land and animals!"

Shivo: "That's crazy! How would the animal even know who the owner was?"

"We will talk about that another day. The point is that people would use money so that they could be the owner of something, and people really liked the idea of having more and more things. There was a whole system built around it, called capitalism, and the idea was that you would try to get as much money as possible, and then you could buy as many things as you wanted. But in the Anti-Miswanting Movement, the AM Movement, people started to advocate for change. Just like nowadays, what people really wanted was to be happy, but back in the day, people thought they should buy and own things to achieve happiness. It took quite some time for the movement to take off, but it was based on the idea that we spent a lot of time and energy working hard to earn the wrong things. Things that didn't make us happy. This excessive work was not just for basic necessities, but a deep desire for unbelievable excess. Everyone was lost in the competition to make more and more and more money to buy nonsense for themselves to own, that did little to make them happy, and caused a lot of poverty and environmental destruction for others."

"The AM Movement started to re-prioritise things. First, it started as small groups and initiatives, but slowly they came together to build Greeniba, the first Untraceable City, do you remember the story?"

Bunkorbio: "Yes! That was the first city of the AM Movement, on an island! We went there on a virtual tour in History Class last year!"

"That's right. There were lots of problems with the old system, and in order to show a drastically different model can work, the AM Movement set up a city on an island off the coast of what was then called Madagascar. And they did *everything* differently. They started by abolishing the concept of money and exchange-based trade. In the old model, everybody was working for themselves, so that they could own things and be better off than others. In the new gift economy model, everybody started working on a single goal, together. Do you remember what it is?"

Shivo: "To live lightly, to restore jointly, and to care for each other. To stick to the triple 95."

"There we go. That's the slogan they use these days isn't it? Back then things started out with similar ideas, although there have been changes along the way. But what really mattered to them was that they set up a city in which people's primary goal was to work together to care better for each other and restore the natural environment around them. Before that time, humankind saw nature as being separate from people, and so much harm was done in just a few centuries. But in Greeniba, everybody would try and contribute to the cause of restoring the environment. It was not about making your own money at the expense of others or nature anymore, but about building a joint sustainable future that prioritised what really makes people happy: having your basic necessities met; belonging and social connection; time affluence; and a healthy body and environment."

Shivo: "But gran, what was the island like? What did they do exactly?"

"Well, I was living on the other side of the world and I never visited, but I followed them in the news. In the beginning, it was tough. They tried to live by the triple 95 rule that we have today. Their goal was to have a city that is 95% compostable, 95% vegetated, and a 95% happiness level among the residents. That was very different from how cities were run, back in the day. So they had to come up with all these innovations, designs, and ways of living that many people thought were crazy. In the beginning they all lived in huts and tents. A big part of the island was previously destroyed by mining activities, and they all worked on restoring this. Planting indigenous plants, rehabilitating rivers, and bringing back wildlife. They had very little beyond the basic comfort, but their population grew from a few thousand to tens of thousands of people, all working together. In terms of ownership, people had very few things, but they were so much happier than people on the mainland. When the rivers started flowing again, even more people migrated there because on the mainland so many rivers were dammed up and polluted that only the wealthy could afford clean water. Working together, they set up a city that had a lot of the infrastructure we now take for granted, like our bamboo skyscrapers and conveyor-belt delivery systems, but those were very new in that time."

Bunkorbio: "Sweet. But tell us about the juicy stuff. When did the money and sex slaves break free? How did they get to the island? And what happened to the slave-masters?

"Bunki! Tuh! So dramatic. Again, the money slaves were from a whole century or two earlier than my life but you are right in that the legacy of slavery lived on for many many decades after slavery was made illegal. The descendants of freedom fighters did really well for themselves, but many others were still poor when I was born. The slave-masters and the freedom fighters remained in power till the very end and were the strongest opponents to radical change. It took a very long time for 'white slave-master men' to fade from the centre of power. They even tried to buy the first island and privatise it into a luxury eco-city for the elite. One Untraceable City insider sold the City out and almost undid the whole project in 2043, but after a decade of fighting a drought, followed by a famine, hit the region. The mainland was starving and the rapid pulling together of Untraceable City, in an impressive way, caused widespread protests across multiple cities whose governments responded slowly and hampered self-organising hunger relief programmes. People refused to fight and insisted that the government implement more substantial environmental and social changes. They promised they would but never did. A few elections later and the political landscape looked very different. Because women presidents were far better at dealing with the Corona Virus of 2019 and the Corona Virus of 2028, women were strongly marketed as the presidents of the twenty-first century and multiple countries had successive waves of female leaders. It wasn't immediate but

174

after a while gender-based issues finally started being addressed. It helped a lot that on Untraceable City Island it was mandatory that the first 10 decades the leaders had to be black women. The world saw how misleading stereotypes were and largely shifted to a new idea of what a leader was."

Shivo: "And why did they stop using planes? Wasn't it great to be in a plane?"

"You have no idea! I was on a plane three times when I was a young woman. It was indeed amazing to fly so fast. But it was also very crowded, and loud, and uncomfortable. People with lots of the special paper, with lots of money, they would travel in more comfort, but the rest of us we sat so close to each other, you have no idea! And I was very lucky, because in fact most people from our half of the planet did not even fly once in their life, because it was very expensive. You know that there were people at the airport who would clean the planes, they would go in them every day, but they were not allowed to stay on the plane when it left? Remember how unequal I told you these times were? We all thought that was normal."

"Airplanes were just like cars. They offered broken promises. When most people needed them the most for getting to work and back fast, there were so many partially full cars on the road that it was incredibly slow. It was quicker to cycle. But people still went by car. We thought that they would give people more freedom, and that people who could use them would be happier. But it was another great Miswanting. Many people would just want to have one car, and then another. Or always take the plane to go far away. The more people got, the more they wanted."

Bunkorbio: "So do you really think our platforms and our helioships are better than cars and planes gran? They are so slow!"

"You kids always think that going fast is best. It was the same when men were in power. But you have no idea what price we paid for that. So much pollution. So much inequality. So little social engagement. When people took the car, they often sat in the car by themselves. Even in the planes, people wouldn't form lasting friendships. Think about that! Sure, when you take the platform, you have to pedal yourself. But you have 5 or 6 other people around you doing the same. And isn't that where you met Orkazio and Roxia? It's so common to make friends that way. And imagine, a trip that took a few hours on a plane now takes a week by helioship. You still get to fly, and you can still visit faraway places. But there is nothing wrong with taking some time. If it weren't for helioships, I would have never met your grandfather!"

"We still have challenges in our society, and I'm not saying it's perfect. Now that we are globally settled into a new era, some people are pushing for subtle forms of excess, striving and elitism that we fought for decades to iron out. It is important to remember our history and avoid the Great Miswanting and The Great Rush from ever happening again. We must teach the next generation or face auto-environmental-destruction yet again. Nowadays, the world is slower and gentler for everybody in it, and that is something worth fighting for."

Biography: Frances first qualified in Zoology and Ecology and later Environment, Sustainability and Society. Her interests sit at the point where humans and nature meet, and how that interaction is negotiated. The is co-founder and director of Communitree, an urban greening and restoration non-profit operating in Cape Town, South Africa.

Paul started his career in web development before moving to industrial ecology. He is currently the Executive Director of the International Society for Industrial Ecology, and is co-founder of an open source urban metabolism collective called Metabolism of Cities. Paul grew up in The Netherlands but spent his adult life in Nicaragua and South Africa.

— Thoughts on Urban Infrastructure —
2100–a lifetime or so away

Thomas L. Theis[+]

In 1962 I was a sophomore in high school. That year Rachel Carson's *Silent Spring* was published (Houghton Mifflin). This, plus a series of well-publicized environmental disasters, elevated human concerns about the quality of the environment with parallel calls for action. A mere eight years later the National Environmental Policy Act in 1970 (NEPA) was passed and the US Environmental Protection Agency was formed. What followed was a flurry of new national and state laws passed throughout the 1970s and 80s aimed at environmental protection. NEPA itself "…*declare[d] national policy which will encourage productive and enjoyable harmony between [humankind] and [its] environment; to promote efforts which will prevent or eliminate damage to the environment and biosphere and stimulate the health and welfare of [humankind]; to enrich the understanding of the ecological systems and natural resources important to the Nation…*". Lofty goals. Were any of them attained? Well, sort of. It was clear that waste discharges were a threat to human health and had to be controlled and to a large extent they were, at least the most visible ones. As importantly, modern industrial societies came to the expectation that a clean environment is a human right.

In 1972, the year I received my Ph.D., another book titled *Should Trees Have Standing?* was published (Oxford University Press). The author was Christopher D. Stone, an environmental lawyer then teaching at the University of Southern California law school. In the book, Stone explored the basis for the environment having legal standing, much like a human plaintiff or victim. He chose trees as his metaphorical example since issues surrounding forestry (uses, value, sustainability, protection afforded to many other species) were much in the news then, but the core of the book extended to the environment more generally. It was quite controversial in its day (and today too—it's now in its third edition). This was during the second period of environmental awareness in US history (the first occurred during the mid-nineteenth century, and we are

[+]University of Illinois at Chicago | theist@uic.edu

now enmeshed in the third). Although my course of study was influenced by events of the day, in truth, I didn't pay much attention to the book. That is, until I started teaching a couple of years later when I found myself involved in a new graduate training program in environmental law and engineering. The program didn't advocate for actual environmental standing, but we did use engineering principles and available laws to devise a series of legal strategies to combat contamination of environmental resources, and force remediation of those that were already contaminated.

Today, we have yet to get to the point where trees, fauna, flora, the atmosphere, lithosphere, and aquasphere can directly take us to court, but progress has certainly occurred for more sophisticated frameworks from which to defend the environment. As far back as 1970, the concept of *ecosystem services*, benefits that well-functioning ecosystems provide to humans at little or no cost, was introduced (e.g. *Study of Critical Environmental Problems*: *Man's Impact on the Global Environment* (MIT Press)). These were further explored and popularized during the Millennium Summit of the United Nations in 2000 and are key to understanding several of the Sustainable Development Goals (SDGs), introduced in 2015 and which we are urged to be realize by 2030. The SDGs are undergirded by the recognition that the earth, today, functions through tightly coupled human-techno-socio-ecological systems. This is especially critical for urban systems because of their population density. Essentially all the SDGs rely upon engineering analysis and design to realize progress, yet for all their trenchant value they provide little guidance on the engineering paradigm needed. And let's be honest, these "services" were around long before humankind existed. We have anthropomorphized them so we can understand and use them in various way. I suppose that's an improvement on the times when they were treated as obstacles to be ignored or overcome.

The *Livable Cities* project at the University of Birmingham (UK) has recently published *The Little Book of…* series, each focused on a different part of improving city life. Among these *The Little Book of Ecosystem Services in the City* was published in 2018 (Sadler, Grayson, Hale, Locret-Collet, Hunt, Bouch, and Rogers, ImaginationLancaster). The authors provide a conceptual basis for engineering analysis and design in facilitating "green" infrastructures. The systems(s) involved can be represented as a set of stock and flow relationships in which natural capital, urban green infrastructural systems, and ecosystem functions interact to deliver ecosystem services in the most efficient manner possible. In this context, urban green infrastructures deliver services while protecting ecosystem function and conserving natural capital within acceptable

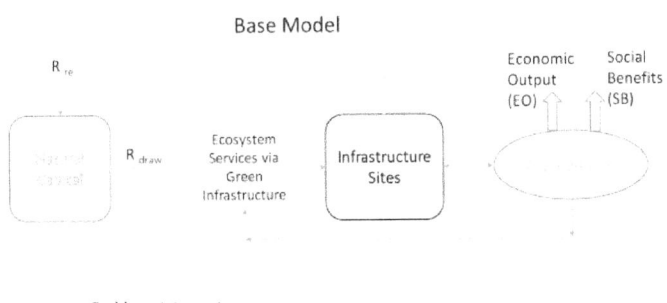

Base Model

Problem statement:

Max : NPV (EO+SB), subject to physical and cost constraints,

and such that $\overline{R}_{re} \geq \overline{R}_{draw}$

limits. This, it seems to me, is entirely consistent with the goals of NEPA. In engineering balance form, the rates of natural capital depletion should not exceed the rates of replenishment, at least over the long term. Diagrammatically the base engineering model might be depicted as:

where the problem statement seeks to maximize present net value of economic and social benefits while conserving time-averaged natural capital. This is a start. At the basic engineering-science level, quantitative expressions of these concepts require considerable inquiry into how ecosystem functions work, the most appropriate measurements for characterizing system attributes, specific ways of expressing system constraints, and how stakeholders perceive, use, adapt, and interact with the coupled system. And technological innovations are necessary to apply this new understanding in the context of urban infrastructure. Simple, huh? No, of course not.

There isn't anything magical about the year 2100. None of the authors in this collection are likely to be around then. Eighty years is more-or-less a human lifetime, and whatever progress is made toward balanced urban systems during that time, it will come incrementally—increments that I like to think have already begun.

Biography: Professor Theis is Director of the Institute for Environmental Science and Policy (IESP) at the University of Illinois at Chicago. IESP focuses on the development of new cross-disciplinary research initiatives in the environmental area. His areas of expertise include the mathematical modeling and systems analysis of environmental processes, industrial pollution prevention, industrial ecology, material flow analysis and life cycle assessment, the environmental chemistry of trace organic and inorganic substances, interfacial reactions, subsurface contaminant transport, and hazardous waste management.He is the founding director of the Environmental Manufacturing Management Program, and NSF IGERT site.

— Crab Air Purifier Vehicle and Octopus Water Treatment Machine —

Truong Thi My Thanh[†]

It was a fresh morning of the first day of new year, welcoming the very good 2100. In a house on Hoang Cau Street in Hanoi, the young mother hurriedly prepares a meal for the New Year. 11-year-old son and 8-year-old daughter chirping beside her, they were so happy after New Year's Eve full of joy. Last night, two little kids with their parents watched fireworks to welcome the new year, the biggest fireworks party ever.

It is lightly raining outside, typical for the tropical spring weather of the North. On the porch, the young father called his son "Big boy, go and operate the air cleaner for me. At the beginning of the year, we should have a very clean environment and hope that all our neighborhoods will always breathe fresh air". The cleaning-air car was designed by both father and son. After having the design idea, they bought components for assembling. Hustling for over two months, they finished the Crab Air Purifier Vehicle. The air purifier works very well. Since the day the Crab Air Purifier Vehicle has been created, the whole neighborhood has been breathing fresh air.

This car has a crab shape. There are pincers on the crab-vehicle to suck polluted air. Air is put into the central part to assess the level of pollution, then purified. Finally, the pincers spray clean air back into the environment. The idea is illustrated on Picture 1, a reflection by an 11-year-boy in Hanoi, Vietnam. He dreams about a clean and sustainable infrastructure in Hanoi, based on the help of "Protect our air environment" vehicles which is a big-scale air purifier.

[†]University of Transport Technology | thanhttm@utt.edu.vn

Picture 1: Crab Air Purifier Vehicle. Author: Pham Nhat Minh, 11 years old - Doan Thi Diem Secondary School, Hanoi, Vietnam

There is a lake called Hoang Cau Lake next to the house with two little kids. It is a manually dugged lake with green surrounding landscape but having heavily polluted water recently. Linh, a 8-year-old girl, would observe the white-bellied fish and find they have difficulty in breathing. She loved the small animals and fishes around her. Therefore, the sight of the dead fish made her think forever. One afternoon, she cheerfully showed her parents the idea of a special octopus that knows how to clean the lake water. The idea was taken from the divine tentacles of the giant octopus, working together smoothly. Four tentacles suck dirty water from the tank to the central body to treat and clean. After the water was treated thoroughly, the remaining four tentacles discharged clean water to the lake. The machine operates with energy generated from a solar battery system on the head, which are designed into the crab's eyes.

The father and kids, eager to study water purification formulas, purchased all the devices need on the mechanical market of Vietnam. After four weeks of testing and re-fitting, the shape of the Octopus Water Treatment Machine was completed. The father and kids show the mother excitedly, their faces looking very proud. The first test is taken at a corner of the lake. The first three days, the machine malfunctioned significantly multiple times, resulting in the team adjusting each component. By the fourth day on, the giant octopus was operating smoothly. After two weeks, the water on lake was much cleaner, becoming transparent to be seen the bottom. The crabs and fish were swimming healthily. A few shrimp jumped to the surface of the water. On the edge of the lake, a few small crabs are crawling. Linh is the happiest person. From now on, she knows how to save the cute creatures in the lake. The idea is illustrated on the Picture 2 which describes a reflection of an 8-year-old girl from Hanoi, Vietnam.

Picture 2: Octopus Water Treatment Machine. Author: Pham Nhat Linh, 8 years old - Doan Thi Diem Primary School, Hanoi, Vietnam

Biography: Truong Thi My Thanh is a Lecturer in the Transport Economics Department at the University of Transport Technology, Vietnam. Her research interest focuses on transport planning and transport economics. Further, her research is based on extensive investigation of transport problems and issues in developing Asian countries to develop innovative policy recommendations, to develop models and tools in conjunction with government bodies as inputs into city- and national-level policies, and derive practical day-to-day improvement for urban transport systems.

— Healthy Transportation Infrastructure Equals Healthy People —

Josias Zietsman[†]

When I think of transportation infrastructure in the year 2100, two thoughts come to mind. Firstly, it will still serve the same basic needs (moving people and goods) but will use very different, very advanced technologies. The second is more of a question: *Will the future infrastructure truly address human needs, or will we perpetuate current trends of reduced quality of life and greater social inequalities?* Since transportation is so inextricably tied to how we live our lives, it is imperative that we develop a transportation system that promotes both equity and well-being into the future. While it is easy to talk about these goals, it is important to address them through an appropriate framework or lens. This will allow everyone to be focused on the same goals when we have these discussions.

We should first ask what framework should inform our planning and resource allocation for the infrastructure of the future. Many prevailing frameworks exist, and all are good candidates. Examples include sustainability, resiliency, livability, health, etc. All have pros and cons, they often overlap, and they are the source of several debates on their appropriateness for addressing future infrastructure. The key principles coming out of these frameworks deal with issues such as *meeting human needs, improving quality of life, and promoting general well-being.*

The framework dealing with human health, however, has the one element that everybody agrees on – we need to *improve the health of society* and that will result in addressing the broad goals of meeting human needs, improving quality of life, and promoting general well-being. Therefore, I believe health is the framework of the future for addressing transportation infrastructure. Such a framework should evaluate how transportation infrastructure in the future could support or detract from health and quality of life. What does it mean to improve the health of individuals? Broadly speaking, it comes down to addressing two important outcomes – reduce premature mortality (help people

---

[†]Texas A&M Transportation Institute | j-zietsman@tti.tamu.edu

185

to live longer) and reduce morbidity (reduce the frequency and severity of illnesses). We can all likely agree that having a long life while suffering multiple and chronic illnesses does not equate to a good quality of life. A focus on addressing morbidity is therefore a more prudent approach since that will indirectly address premature mortality.

My wife (an art teacher) and my daughter (a medical student) made the sketch below, to help me illustrate how I view the link between health and transportation and the importance of health as a framework to make good decisions regarding transportation infrastructure. It shows an evolution from the current transportation system to one in 2100. It shows a person (representing a proportion of the human race) stepping away from nature (representing the environment and the ecological system) for personal and economic gain. The net effect of this is very short-term gains but long-term pain inflicted on the human race. We also see a young medical doctor in deep concentration holding her stethoscope over the transportation system evaluating its health and the resulting health of the population. The final part of the sketch indicates a bright future with a good balance between the future transportation system and the environment, showing that a good and healthy outcome is possible.

Research has shown that clinical care (working directly with patients) addresses only 20% of health outcomes, while the remaining 80% can be attributed to behavior, socio economic issues, and the physical environment. The majority of health outcomes can therefore be addressed upstream of clinical care. Transportation infrastructure is clearly an upstream investment. This investment in transportation infrastructure in the United States is currently valued at close to $8 trillion with a tremendous impact on the health outcomes of individuals, whether from exposure to emissions, the ability (or inability) to use active transportation, and access to health-promoting destinations. A shift in focus to a more health-related framework will also direct the focus on how infrastructure investments create the best outcomes.

The key question is, how would we evaluate the health of the transportation system so that it can keep us healthy and improve our quality of life? Transportation results in both positive and negative health impacts through various mechanisms. My colleagues and I have identified 14 such mechanisms, termed as pathways. Four of these pathways have beneficial health impacts and 10 have detrimental health impacts. The four with beneficial impacts include: green space, physical activity, access, and mobility independence. The 10 with detrimental impacts include: contamination, social exclusion, noise, urban heat islands, vehicle crashes, air pollution, community severance, electromagnetic fields, stress, and greenhouse gas emissions. Future transportation infrastructure decisions can be better informed by assessing their impact on these 14 pathways.

As the transportation system evolves, we can still use these pathways (or an evolution of these pathways) to better address the impacts of the future transportation system. For example, the framework can be used to evaluate the possible health impacts of a future technology such as autonomous vehicles. Revolutionary modeling techniques using approaches such as big data and machine learning will make it possible to quantify the health impacts in a seamless manner, based on the 14 pathways framework.

In conclusion, if we perpetuate current trends in transportation infrastructure development, we will not be on a sustainable path to the year 2100. The important principles to consider in evaluating our trajectory include meeting human needs, improving quality of life, and promoting general well-being for individuals. Transportation infrastructure can play a major part in supporting these outcomes. The most appropriate framework for evaluating transportation infrastructure related to these goals is a framework developed around the principle of improving human health. To this end, I propose viewing the linkages between health and transportation through the lens of the 14 pathways mentioned earlier. By predicting and quantifying outcomes in this way, we can assess whether our decisions, related to transportation infrastructure, are moving us farther away from, or closer to, a more healthy society now and the year 2100.

Biography: Dr. Zietsman is an Assistant Agency Director and Strategic Advisor at the Texas A&M Transportation Institute. He is also the director of the Center for Advancing Research in Transportation Emissions Energy and Health (CARTEEH). He holds a PhD in Civil Engineering from Texas A&M University and he is a member of the Graduate Faculty of Texas A&M University. His research interests are in sustainable transportation, air quality, public health, and performance measurement.

Index